Cap. 1

DATE		

FORM 125 M

THE
ILLUSTRATED
ENCYCLOPEDIA
OF GENERAL
AVIATION

No. 2274
$14.95

THE ILLUSTRATED ENCYCLOPEDIA OF GENERAL AVIATION

BY PAUL GARRISON

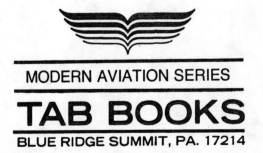

MODERN AVIATION SERIES

TAB BOOKS

BLUE RIDGE SUMMIT, PA. 17214

FIRST EDITION

FIRST PRINTING—NOVEMBER 1979

Copyright © 1979 by TAB BOOKS

Printed in the United States of America

Library of Congress Cataloging in Publication Data

Garrison, Paul.
    The illustrated encyclopedia of general aviation.

    Includes index.
        1. Aeronautics—Dictionaries. 2. Private flying—Dictionaries.
I. Title.
TL509.G27   629.13'003   79-17267
ISBN 0-8306-9754-3
ISBN 0-8306-2274-8 pbk.

Preface

In this illustrated encyclopedia/dictionary we have attempted to alphabetically list and explain every concept, expression, phrase and abbreviation commonly used in relation to general aviation. Included, among other subjects, are all major manufacturers of aircraft and avionics and all of their more important products. Also included are the important aviation organizations and publications, their addresses to be found under *Aviation Organizations* and *Aviation Publications*.

Not included is information related strictly to military aviation and to the airlines.

Paul Garrison

Contents

A—Absolute (temperature).

A—Alaskan Standard Time.

A—Alpha (phonetic alphabet).

A—Alternate (on instrument approach charts).

A—Altimeter (on radio logs).

A—Arctic air mass.

A—Cleared to the airport (ATC clearance shorthand).

A—Ceiling as reported by an aircraft (in sequence reports).

A—Hail (in sequence reports).

AAA—Antique Aircraft Association.

AAAE—American Association of Airport Executives.

AAC—Alaskan Air Command.

AACA—Alaskan Air Carriers Association.

AAP—Advise if able to proceed.

AAP—Association of Aviation Psychologists.

AAS—Airport advisory service.

AAS—Airport advisory station.

AATM—At all times.

AAWF—Auxiliary aviation weather facility.

AB—Air base.

AB—Airborne.

AB—Continuous automatic transcribed broadcast service (AIM).

ABAC—Association of Balloon and Airship Constructors.

ABCST—Automatic broadcast.

ABD—Aboard.

Abeam—In aviation-radio phraseology a term used to describe the position of an aircraft relative to a fixed point on the ground, such as: *Abeam the tower.*

ABM—Abeam.

Abort—The act of terminating a takeoff or other planned maneuver, or the command to do so.

Absolute altitude—Actual height above terrain in feet agl.

Absolute ceiling—The maximum altitude above sea level to which a particular aircraft can climb and then maintain horizontal flight under standard atmospheric conditions.

Absolute humidity—The weight of water vapor per unit volume of air.

ABV—Above.

AC—Advisory circular.

AC—Alternating current.

AC—Altocumulus.

AC—Approach control.

ACAS—Automatic collision-avoidance system.

Accelerated stall—A high-speed stall caused when excessive g-loads are applied quite rapidly.

Accelerate-stop distance—The distance required to accelerate an aircraft from a standing start to a specific speed, usually liftoff speed, and, assuming failure of the critical engine at the instant that speed is reached, to bring the aircraft to a stop, using heavy breaking. The accelerate-stop distance for any aircraft varies drastically depending on runway-surface conditions, wind, temperature and other factors.

Acceleration—Increase in velocity or the rate at which such increase takes place.

Acceleration error—The error in the reading of the magnetic compass which is caused when the compass card tilts in its mounting as a result of changes in the speed of the aircraft. It is especially noticeable on east or

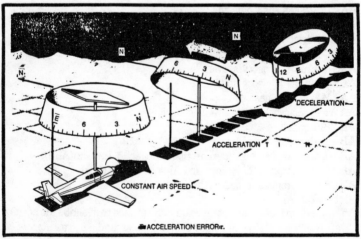

Acceleration error (courtesy of FAA).

west headings. It turns to the north when the aircraft is accelerating and to the south when decelerating.

Accelerometer—An instrument capable of recording the rate of acceleration. Not usually found in light aircraft.

Accident—With relation to aviation the term accident means an occurrence associated with the operation of an aircraft which takes place between the time any person boards the aircraft with the intention of flight until such time when all such persons have disembarked, in which any person suffers death or serious injury as a result of being in or upon that aircraft or by direct contact with the aircraft or anything attached to it. In this context a *fatal injury* is one which results in death within seven days. A *serious injury* means an injury which requires hospitalization for more than 48 hours within seven days from the date on which the injury was received; results in the fracture of any bone other than fingers, toes or nose; involves lacerations which cause severe hemorrhages, nerve muscle or tendon damage; involves injury to an internal organ; or involves second or third-degree burns or any burns affecting more than five percent of the body surface. *Substantial damage* means damage or structural failure which adversely affects the structural strength, performance or flight characteristics of the aircraft and which would normally require major repair or replacement of the affected component. Engine failure limited to an engine, bent fairings or cowling, dented skin, ground damage to rotor or propeller blades, damage to landing gear, wheels, tires, flaps, engine accessories, brakes or wing tips are not considered substantial damage according to FAR Part 430.

Accuracy landing—See spot landing.

ACDNT—Accident.

ACFT—Aircraft.

ACK—Acknowledge.

Acknowledge—A request for verification that the message has been received and understood.

ACLD—Above clouds.

ACPT—Accept.

Acrobatics—See aerobatics.

ACSL—Standing lenticular alticumulus.

Active runway—The runway in use under prevailing wind, weather and/or noise-abatement conditions.

ACTV—Active.

Adams Balloon Loft, Inc.—Mike Adams Balloon Loft, manufacturers of a variety of hot-air balloons. (P.O. Box 12168, Atlanta, GA 30355).

ADAP—Airport Development Assistance Fund.

ADC—Air-data computer. A digital computer using input from air-data instruments in altitude and speed-control alert functions.

ADF—Automatic direction finder.

ADF approach—A non-precision instrument approach using a non-directional beacon, an LF/MF nav aid or standard broadcast station as an aid in navigational guidance. (See also *non-precision approach*).

ADI—Attitude director indicator. The visual display associated with flight directors. (See also *flight directors*.)

Adiabatic lapse rate—The rate at which the temperature of the air changes without any outside source for this change being involved. Temperature increases when air is compressed and it decreases when air rises and expands. When the air is dry this change occurs at a rate of 5.5 degrees F. per 1,000 feet of change in altitude. With moist air, condensing water vapor gives off heat and at the saturation point the lapse rate is three degrees F. per 1,000 feet of increase in altitude.

ADIZ—Air Defense Identification Zone.

ADJ—Adjacent.

Adjustable-pitch propeller—A propeller on which the blade angle can be changed on the ground to achieve the most efficient operation for special conditions, such as flight at high altitudes.

ADMA—Aviation Distributors and Manufacturers Association.

ADMIN—Administration.

Administrator—The Administrator of the Federal Aviation Administration. Also refers to any person to whom he has delegated appropriate authority in a particular situation.

AD Notes—Advisories issued by the FAA, usually as a result of the detection of some type of structural or other defect of an aircraft, after it has been certificated and is in service. Compliance with AD notes within the period of time stated therein is mandatory and the responsibility of the aircraft owner or operator.

ADVC—Advice.

ADVCTN—Advection.

Advise intentions—A phrase used by ATC, requesting the pilot to state what he is planning to do.

Advisory—Information provided by ground facilities or pilots which is of a nature as to be helpful in assisting pilots in the safe and efficient conduct of a flight; such as weather, airport conditions, traffic, out-of-service navigational facilities, etc.

Advisory for light aircraft—(AIRMET)—An in-flight weather advisory covering weather phenomena of importance to light aircraft and of potential importance of all aircraft; such as moderate icing, wide-spread but moderate turbulence, visibilities less than two miles, ceilings less than 1,000 feet, and winds of 40 or more knots near the surface.

ADVN—Advance.

ADVY—Advisory.

ADZ—Advise.

AEA—Aircraft Electronics Association.

AERO—Aeronautical or aeronautics.

Aero Commander (courtesy of Rockwell International).

Aero—A monthly aviation magazine sent free of charge to some 90,000 aircraft owners in the U.S.

Aerobatics—Maneuvers involving abrupt changes in altitude and often resulting in unusual attitudes not used in normal flight. These may include all manner of stunts. Aircraft designed for this type of use must be especially stressed, normally to a minimum of six positive and three negative g. Also referred to as *acrobatics*.

Aero Commander—One of the first piston-twin aircraft designed specifically for business and corporate use. Designed by Ted Smith, it was flown coast to coast with one propeller removed to prove its single-engine reliability. The original manufacturer was subsequently bought by Rockwell International (then North American Rockwell), which continues to produce an entire family of Commander aircraft, most based largely on the original design.

Aerodrome—Airport. A term used by most ICAO member nations.

Aerodynamics—The forces, such as resistence, pressure, velocity, and others involved in the movement of air or gases around a moving body. Conversely, the branch of dynamics and physics dealing with these forces.

Aero Mechanism—Manufacturers of encoding altimeters and related products. (7750 Burnet Avenue, Van Nuys, CA 91405).

Aeronaut—A pilot. The term is used practically exclusively in ballooning.

Aeronautical advisory station—See Unicom.

Aeronautical beacons—Displays of stationary or rotating white and/or colored lights indicating an airport, the route leading to an airport or a point from which bearings can be taken to an airport, an outstanding landmark or ground-based hazard.

Aeronautical charts—A map developed especially for use in aviation navigation. The U.S. government periodically publishes a variety of such charts and pilots are urged to operate with the latest editions. These charts include 55 Sectionals at a scale of 1:500,000, covering the 48 contiguous United States, Hawaii, Alaska and portions of the Caribbean, northern Mexico and southern Canada. World Aeronautical Charts called WAC's or ONC's are produced at a scale of 1:1,000,000 and are available for every land portion of the free world. There are 57 of those. Then there is one VFR/IFR planning chart covering the 48 contiguous states. In addition there are three jet navigation charts (JNC's) covering all of Northern America. For IFR operations a variety of radio-navigation charts are published, both, by the government and by the Jeppesen-Sanderson. These are low-altitide en-route charts (government: 28, Jeppesen: 32 for the contiguous states) showing primarily the airways, distances between nav aids, VORs, VORTACs, NDBs LF/MF nav aids, airports of importance, MEAs, MRAs, nav and com frequencies and other information applicable to operating within the ATC system. There is also a nearly indefinite number of approach charts picturing and describing the authorized instrument approaches to various airports. Then there are so-called SID and STAR charts showing standard instrument departure and arrival procedures. There are special large-scale charts for each of the TCAs. In addition, many states publish aeronautical charts of the state, some available free of charge, some for a small fee. Charts are available from the U.S. Department of Commerce, National Oceanic and Atmospheric Administration, National Ocean Survey, Washington, D.C. or from the charts department of the AOPA and NPA (at no increase in price). Charts produced by Jeppesen-Sanderson can be ordered on a subscription basis from Jeppesen-Sanderson, Inc., 8025 East 40th Avenue, Denver, CO 80207.

Aeronautical planning chart—A chart covering the 48 contiguous United States, designed for use in flight planning. Scale: Approximately 70 nm per inch.

Aerosonic Corporation—Manufacturers of encoding altimeters and related equipment. (3312 Wiley Post Road, Carrollton, TX 75006).

Aerospatiale Helicopter Division—U.S. representative and subsidiary of the French manufacturer Aerospatiale, manufacturers of a variety of

AStar (courtesy of Aerospatiale).

high-performance helicopters; AStar 350, a five-seat single engine turbine helicopter; Gazelle, also a five-seat single-engine turbine helicopter;
Alouette III, a seven-seat single-engine turbine helicopter; Dauphin,
eight-seat single- and twin-turbine helicopters; Puma, a 16-seat twin-
turbine helicopter, and the Lama, a high-performance utility helicopter.

Aerostar—A high-performance piston twin aircraft of the owner flown
class, available with or without turbocharging. Designed by Ted Smith
and originally marketed by his own company, the manufacturing and
marketing rights have been acquired by Piper Aircraft Corporation which
has renamed the aircraft the Piper Aerostar. It is the fastest piston twin
currently in production. Seating capacity: Six.

Aerostat—A hot-air or gas balloon.

Aerostation—Same as aerostat.

AF—Air Force.

AFB—Air Force base.

Affirmative—In aviation-radio phraseology the term is used to indicate
correct or *yes*.

AFHF—Air Force Historical Foundation.

AFMF—Air Force Museum Foundation.

AFT—After.

Aloutte III (courtesy of Aerospatiale).

Aerospatiale turbine helicopter specifications.

AEROSPATIALE HELICOPTER CORPORATION	ASTAR AS 350D	GAZELLE (stretched) SA 341G	LAMA SA 315B	ALOUETTE III SA 319B	DAUPHIN SA 360C	PUMA S 330J	TWINSTAR AS 355E	DAUPHIN 2 SA 365C
ENGINE	L/LTS 101-600A2	Astazou IIIA	Artouste IIIA	Astazou XIVB	Astazou XVIIIA	Turmo IVC (2)	Allison 250-C20F (2)	Ariel
shp	592	615	562	592	871	1.558 each	420 each	642 each
TBO (hours)	2.400	1.500	1.750	1.500	1.500	1.750	3.000	2.000
SEATS	6	5	5	7	10 (max 14)	22	6	10 (14 max)
RATE OF CLIMB, sea level (fpm)	1.713	1.338	1.083	886	1.400	1.200	NA	2.010 (1 eng. 492)
SERVICE CEILING (ft)	15.000	14.925	17.720	13.450	14.270	15.750 (1 eng. 7.220)	7.218 (preliminary)	15.910 (1 eng. 3.935)
HIGE (ft)	10.000	9.185	16.565	10.170	8.035	7.545	4.922 (preliminary)	6.690
HOGE (ft)	7.380	7.215	15.100	5.575	5.740	5.580	NA	2.790
MAX SPEED, sea level (knots)	147	168	113	118	170	142	NA	170
NORMAL CRUISE SPEED (knots)	128	142 (economy:125)	103	105	146 (economy:133)	134	129 (economy 119)	136
Vne (knots)	147	168	113	118	170	142	NA	170
FUEL CAPACITY (lbs)	938	804	1.018	1.018	1.132	2.747	1.293	1.132
RANGE w. full fuel (nm)	427	346	278	340	353	NA	NA	245
w. max payload (nm)	345	346	278	NA	324	NA	NA	245
GROSS WEIGHT (lbs)	4.190	3.970	4.300 (extern ops 5.070)	4.960	6.615	16.317	4.630 (extern ops 5.072)	7.495
EMPTY WEIGHT (lbs)	2.360	2.127	2.276 (w sling 2.311)	2.557	3.797	8.358	2.712 (w sling 2.752)	4.141
PAYLOAD w. full fuel (lbs)	892	1.039	1.006	1.385	1.686	5.212	625	2.222
PAYLOAD w. fuel for 100 nm (lbs)	1.610	1.610	1.649	2.095	2.495	7.012	1.692	2.892
FUEL w. full payload (lbs)	660	804	1.018	1.018	1.048	2.747	748	1.132
LENGTH/HEIGHT/WIDTH external (ft)	35.7 9.10 12.6 88	31.9 10.45 6.6	33.58 10.14 7.8	33.38 9.84 8.54	36.02 11.48 7.37	48.62 16.9 9.8	35.7 9.9 97.6 88	36.02 11.5 10.3
cabin (ft)	7.9 4.3 5.4	5.6 3.52 4.3	6.43 4.35 4.36	5.97 4.36 6.1	7.5 4.6 6.3	19.9 5.1 5.9	7.94 5.4 4.26	7.5 4.6 6.3
MAIN ROTOR number of blades	3	3	3	3	4	4	3	4
diameter (ft)	35.7	34.45	36.15	36.16	37.73	49.2	35.07	38.3
TAIL ROTOR number of blades	2	Fenestron	3	3	Fenestron	5	2	Fenestron
diameter	6.1	2.28	6.27	6.27	3	10	6.1	3
IFR CERTIFICATION STATUS	No	CAT II single pilot	No	No	Single pilot	Two pilots	No	Two pilots
PRICE (1979 $s)	252.000	313.500	408.000	504.000	675.000	2.432.000	510.000 (1981 $s)	995.000
STANDARD EQUIPMENT								
RMI	No	No	No	No	1	2	No	1
AUTOPILOT	No	No	No	No	No	Yes	No	No

In the Engine column manufacturers are indentified as follows:

C/ = Teledyne Continental

GA/ = Garrett AiResearch

GE/ = General Electric

L/ = Avco Lycoming

PW/ = Pratt and Whitney

RR/ = Rolls Royce

In the Propeller column const. speed = constant speed; ff = full feathering.

All *measurements* are in feet and 1/10s of feet. All *speeds* are in knots. All *weights* and all *fuel* figures are in pounds. All *range* figures are in nautical miles.

Payload is calculated as *useful load* minus fuel, and in turboprop and jet aircraft also minus a professional crew of two (400 lbs).

Aerostar 600A (courtesy of Piper).

AFTN—Afternoon.

Ag aviation—Agricultural aviation, such as using aircraft for seeding or pest control.

Ag Cat—A heavy-duty agricultural biplane originally produced by Grumman Aviation and now produced and marketed by Gulfstream-American Corporation.

agl—Above ground level.

Agonic line—The line of zero magnetic variation. In the U.S. it runs approximately southeast from the middle of Lake Superior to Savannah, Georgia.

Agusta 109A—A twin-engine turbine helicopter manufactured by Giovanni Agusta S.p.A. in Italy.

Ag Wagon—A low-wing agricultural aircraft manufactured by Cessna Aircraft Company.

AHD—ahead.

Gulfstream-American Ag-Cat (courtesy of Gulfstream-American).

Agusta 109.

AIA—Aerospace Industries Association.

AIAA—American Institute of Aeronautics and Astronautics.

AID—Airport information desk.

Aileron—The primary control surface located at the trailing edges of the outer wing panels which, when moved up or down, causes the airplane in flight to bank. Movement of the control column (wheel, yoke or stick) to the right raises the right aileron and lowers the left, thus inducing a right bank. Some ailerons are referred to as *drooped*, meaning that both can be deflected either downward or upward simultaneously (while maintaining their lateral control function), thus adding a small degree of lift during low-speed operation or, conversely, effecting a slight degree of increase in speed during cruise. On some aircraft the function of the ailerons is accomplished by the use of spoilers. For details see *spoilers.*

Aileron roll—An aerobatic maneuver in which full aileron deflection causes the airplane to rotate (roll) around its lengthwise axis without a change in heading.

AILS—Automatic instrument landing system.

AIM—Airman's Information Manual.

Air—The air we breathe and which permits us to fly consists of 78.09 percent nitrogen, 20.95 percent oxygen plus minute quantities of argon, carbon dioxide, neon, helium, krypton, xenon, ozone and radon, not to mention moisture, dust and a huge variety of chemical pollutants caused by industry, automobiles, aircraft, etc. At sea level the average pressure of a column of air measuring one square inch is about 14.5 pounds, which corresponds to the weight of a comparable column of mercury 29.92 inches high. This then is the origin of our standard altimeter setting.

Airborne telephone systems—These systems permit pilot or passengers to contact any number of ground-based stations and from there to be connected to any telephone anywhere in the world. There is no monthly charge. Charges are a fixed fee per (completed) call plus the applicable long-distance rate from the ground station to the party being called.

Agusta 109 twin-turbine helicopter specifications.

GIOVANNI AGUSTA S.p.A. MODEL:	109A
ENGINES	Allison 250-C20-B
shp each	420
TBO (hours)	3,500
SEATS	8
RATE OF CLIMB sea level (fpm)	1,620 (1 eng. 340)
SERVICE CEILING (ft)	16,300 (1 eng. 7,000)
HIGE (ft)	9,800
HOGE (ft)	6,700
MAX SPEED, sea level, (knots)	168
NORMAL CRUISE SPEED (knots)	147
Vne (knots)	168
FUEL CAPACITY (lbs)	978
RANGE w. full fuel (nm)	365
RANGE w. max payload (nm)	123
GROSS WEIGHT (lbs)	5,730
EMPTY WEIGHT (lbs)	3,800
USEFUL LOAD (lbs)	1,930
PAYLOAD w. full fuel (lbs)	952
w. fuel for 100 nm (lbs)	1,657
FUEL w. max payload (lbs)	330
LENGTH/HEIGHT/WIDTH external (ft)	42.8/10.8/9.4
cabin (ft)	5.3/4.6/4.7
MAIN ROTOR number of blades	4
diameter (ft)	36.1
TAIL ROTOR number of blades	2
diameter (ft)	6.7
IFR CERTIFICATION STATUS	dual/single pilot
PRICE (1979 $s)	797,500

Cessna AG Wagon (courtesy of Cessna).

Airborne telephones comparison chart.

MANUFACTURER	MODEL	PRICE uninstalled	Volts input	Push-to-talk	Duplex	Add'l handsets	Solid state 100%	Units	Weight lbs.	REMARKS
ASTRONAUTICS	SS-11	595	28	■			■	4	29.75	
KING	KT 96	1,255	14 28*					1	3.5	*28V ADAPTER $135.
WULFSBERG	FLITEFONE II SE	2,078	14 28			■		2	7.8	
ASTRONAUTICS	SS-IIIS	2,295	14 28			■		3	11	
ASTRONAUTICS	SS-IIID	2.757	14 28					3	13.5	
WULFSBERG	FLITEFONE III	3,301	28		■			2	9.9	

AIRBORNE TELEPHONES

Persons on the ground can contact the aircraft in flight if they know the number assigned to the particular telephone in the aircraft, and the approximate location of the aircraft at the time the call is being placed. Airborne telephone systems are produced by three manufacturers (see chart).

Airborne weather radar—Radar carried aboard the aircraft and used to detect areas of precipitation, especially in relation to thunderstorms where such areas are indicative of the greatest degree of turbulence. Historically, the radar display has been analog in monochrome (green or black). Recent developments have introduced digital computer technology in airborne radars and most new models are digital, using a computer-generated display which most pilots find easier to interpret.

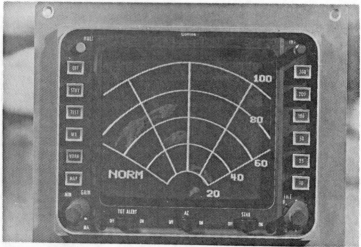

Airborne weather radar (courtesy of Collins).

Airborne weather radar comparison chart.

MANUFACTURER	MODEL	PRICE uninstalled	Volts input DC	Volts input AC 400 Hz	Output kw (peak) amps	Frequency MHz (x band)	Ranges nm 0-50	50-100	100-200	200-300	Scan, degrees	Antenna tilt deg.	Antenna flat plate	Antenna paraboloic	Readout black & white	Readout color	scope, in. dia	Sizes indicator wide/hi/deep	Stabilization pitch degrees +/-	Stabilization roll degrees +/-	Display digital	Display DST	Units	Weight lbs.
KING	KWX 50	6.795	28		3.4	9375	30	60	150		90	+/-12	OPT		●		5	6.25×4 ×11.1	X				3	20.9
RCA AVIONICS	PRIMUS 20	7.235 to 10.845	28		4	9375	10 20 40	80		200	90	+/-15	●		●		5.4	5.25× 5.25×10	X				3	20.9
COLLINS	WXR-150	7.240	28		3	9345	30	60	120		90	+/-15		OPT-T	●		4.3	6.25×4 ×8.3			●		3	14.9
BENDIX AVIONICS	RDR-160	7.735	28		3.5	9375	5 10 20 40	80	160		90	+/-15	●		●		4.3	6.25×4 ×9.9			●		2	18.7
BENDIX AVIONICS	RDR-150	9.913	28		3.5	9375	5 10 20 40	80	160		90	+/-15	●		●		4.3	6.25×4 ×9.9			●		3	23.5
RCA AVIONICS	WXR-200	10.980	28		3.5	9345	30	60	120 180		90	+/-15	●		●		4.3	6.26×4 ×8.3	X		●		3	23.7
BENDIX AVIONICS	RDR-160	11.285	28		3.7	9375	5 10 20 40	80	160		90	+/-15	●		●			6.25×4.7 ×12.06			●		3	24.7
RCA AVIONICS	PRIMUS 21	12.925	28		3.5 / 1	9375	10 20 40	80	120		90	+/-12	●		●		5.4	5.25× 5.25×10	X		●		3	23.7
BENDIX AVIONICS	RDR-150	13.463	28	115	3.5	9375	5 10 20 40	80	160		90	+/-15	●		●		5.4	6.25×4.7 ×12.06			●		3	27.3
RCA AVIONICS	PRIMUS 30A	15.220	28		3.5 / 3	9375	10 20 40	80		200	120	+/-12	●		●		5.4	5.25× 5.05× 10.5	25	25	●		3	23.5
BENDIX AVIONICS	RDR-1100	16.066	28	115	3.75 / 1.15	9375	10 20 40	50	100	200	120	+/-15	●		●		4.3	6.25×4 ×9.9	30	30	●		3	22.2
COLLINS	WXR-250A	16.980	28	115	3.5 / 1	9375	5 10 20 40	80	120	240	120	+/-15	●		●		4.3	5.2×5.2 ×9.8	30	30	●		3	27.3
RCA AVIONICS	PRIMUS 31	7.320	28	115	4 / 35	9345	30		120	200	120	+/-12	●		●	●	5.4	5.25× 5.05× 12.3	25	25	●		3	33.8
RCA AVIONICS	PRIMUS 300	18.870	28	115	3.5 / 3	9375	5 10 20 40	50	100	200	120	+/-15	●		●	●	5.1	6.35× 6.25× 12.3	25	25	●		3	26.5
BENDIX AVIONICS	RDR-1100	19.593	28	115	3.5 / 3	9375	10 20 40	80	160	240	60 120	+/-15	●		●	●	4.3	6.25×4.7 ×12.06	30	30	●		3	27.75
BENDIX AVIONICS	RDR-1200	20.717	28	115	10	9345	5 10 25	50	100	200	120	+/-15	●		●	●	4.3	6.25× 6.09×	30	30	●		3	34.8
RCA AVIONICS	PRIMUS 40	21.000	28	115	10	9345	10 25	50 75	150	200 300	120	+/-15	●		●	●	5.4	6.25× 6.25× 4.75×7	30	30.89	●		3	28.2
COLLINS	WXR-300	22.660	28	115	5	9375	10 25	50 75	150	300	120	+/-15	●		●	●		6.25× 6.25×	30	30	●		3	30.7
BENDIX AVIONICS	RDR-1300	23.209	28	115	10	9375	10 20 40	80	160	240	40 120	+/-15	●		●	●	4.3	6.25× 6.25× 10.875	30.89	30.89	●		3	35.7
BENDIX AVIONICS	RDR-1200	24.500	28	115	10	9375	25.5 10 20 40	80	160	240	60 120	+/-15	●		●	●		6.25× 6.36× 12.5	30	30	●		3	38.1
RCA AVIONICS	PRIMUS 400	25.040	28	115	3.5 / 3	9375	10 25		200	300	120	+/-15	●		●	●	5.1	6.25×10.875 6.25× 6.25×	30.89	30.89	●		3	3
BENDIX AVIONICS	RDR-1300	27.069	28	115	6 / 35	9375	24 10 25	75	160	240	60 120	+/-15	●		●	●		6.25× 6.25×	30.89	30.89	●		3	30.7
RCA AVIONICS	PRIMUS 50	27.320	28	115	5 / 40	9375	10 25		150	240	120	+/-15	●		●	●	5.4	6.25× 6.25×	30	30	●		3	35.9
BENDIX AVIONICS	RDR-1400	29.907	28		3.5 / 3	9375	25.5 10 20 40	50 100	80	300	40 120	+/-15	●		●	●	4.3	6.25× 6.36× 12.5	30.89	30.89	●		3	30.7
RCA AVIONICS	PRIMUS 90	35.315	28	115	5.7 / 35	9345	10 25	50 100		300	160	+/-15	●		●	●	5.1	6.36× 6.36× 12.5	30	30	●		3	50

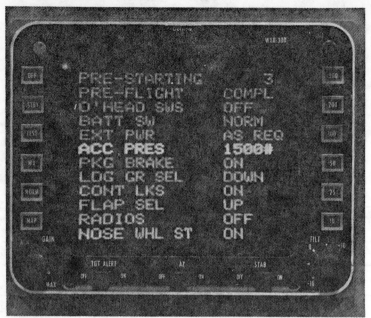

New weather radar models are capable of projecting a variety of information (courtesy of Collins).

Latest models are also capable of projecting suggested flight routes on the radar display and can be used to display checklists and a virtually endless variety of data stored in the computer memory and recalled at command by the pilot, producing an alpha-numeric readout. (See chart).

Air-cooled engine — Term used primarily in relation to piston engines cooled by outside air passing directly over the cylinders, the heat of which is generally transferred to the air by flat cooling fins or baffles.

Aircraft—Any man-made object that flies. Primarily used with reference to fixed-wing aircraft, helicopters, gyroplanes, gliders and sailplanes, hot-air and gas balloons, blimps and dirigibles, but technically also including hang gliders and probably even kites.

Aircraft approach category—Categories based on speeds representing 1.3 V_{so}, or on maximum certificated landing weight. Category A covers aircraft of less than 30,000 pounds and speeds of less than 91 knots. Category B: Weight 30,001 to 60,000 pounds; 91+ knots but less than 121 knots. Category C: Weight 60,001 to 150,000 pounds; speeds 121+ knots but less than 141 knots. Category D: Weight 151,000 pounds and up; speed 141+ knots but less than 166 knots. Category E: Speed over 166 knots regardless of the weight of the aircraft.

Aircraft classes—In order to establish criteria with relation to wake turbulence and the necessary separation minimums, ATC has classified aircraft as follows: *Heavy*: Aircraft capable of takeoff weights of 300,000 pounds or more, regardless of the actual takeoff weight during a particular operation. *Large*: Aircraft with a maximum certificated takeoff weight between 12,500 and 300,000 pounds. *Small*: Aircraft with a maximum certificated takeoff weight of under 12,500 pounds.

Aircraft flight manual— A document issued by the manufacturer which contains certification details, limitations, procedures, performance parameters and all other data and information needed by the pilot for the safe operation of that aircraft. Also known as *owner's manual*.

Aircraft Instrument and Development—Manufacturers of encoding altimeters. (317 East Lewis Street, Wichita, KS 67202).

Aircraft movement—The phrase refers to takeoffs and landings. Aircraft movements are counted by control towers to establish the amount of traffic being handled. Thus two movements constitute one flight of an aircraft. (ARTCCs and FSSs count contacts rather than movements).

Aircraft type designators—The letter- number combinations given to all civilian and military aircraft, such as PA-31T-1 which stands for Piper Aircraft Cheyenne I.

Air Defense Identification Zone— An area of airspace over land or water within which proper identification, location and control of aircraft are required for national security. The zones are principally off the Atlantic, Pacific and Gulf coasts, along the Mexican border, around Hawaii and over most of Alaska. There is also a Distant Early Warning Identification Zone (DEWIZ) located primarily over Alaska and parts of northern Canada. As a general rule the filing of IFR or CVFR flight plans is mandatory before entering this airspace.

Air density—The mass of molecules in a given volume of air. This changes with variations in air pressure. (See *density altitude*).

Air density error—The error in the presentation of an airspeed indication caused by changes in air density resulting from changes in altitude and barometric pressure. True air speed (TAS) can be calculated by using a standard aviation computer.

AiResearch Manufacturing Company (Arizona/California)— Manufacturers of sophisticated avionics systems, of the TFE 731 fan jet engines and the TPE 331 turbine engines for turboprop aircraft, plus a wide variety of related systems. Also, modifiers of aircraft such as the Lockheed JetStar and the British HS 125 series. Also, operators of some fixed-base operations. (A subsidiary of Garrett Corporation. AiReasearch/California, 2525 West 190th Street, Torrance, CA 90509. AiResearch/Arizona, 402 South 36th Street, Phoenix, AZ 85034).

Airfile—Means filing an IFR flight plan while airborne. Rarely used with reference to filing a VFR flight plan from the air.

Airfoil—Any surface designed to create lift, either positive or negative, when moving though the air at a given speed. Primarily, wings and control surfaces, though propellers and helicopter rotorblades are also airfoils.

Air Defense Identification Zones (courtesy of FAA).

20

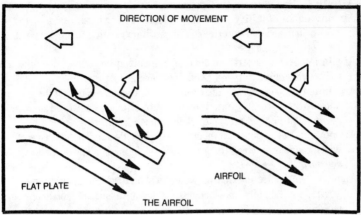

DIRECTION OF MOVEMENT

FLAT PLATE

AIRFOIL

THE AIRFOIL

The Airfoil (courtesy of FAA).

Airframe—The frame, covering and components of an aircraft, not including engine, propellers, tires, avionics and instruments.

Airframe manufacturers—Manufacturers of aircraft: American Jet Industries; Anderson-Greenwood Aviation; Avions Marcel Dassault-Breguet; Beech Aircraft Corporation; Bellanca Aircraft Corporation; Canadair, Ltd.; Cessna Aircraft Company; Foxjet Industries; Gates Learjet Corporation; Gulfstream-American Corporation; Israel Aircraft Industries; LearAvia; Lockhead Georgia Company; Mitsubishi Aircraft International; Mooney Aircraft Corporation; Piper Aircraft Corporation; Rockwell International General Aviation and Sabreliner Divisions; Ted Smith Aerostar Corporation, a subsidiary of Piper Aircraft Corporation; Swearingen Aviation Corporation; Aerospatiale Helicopter Division; Bell Helicopter Textron; Brantly-Hynes Helicopter, Inc.; Enstrom Helicopter Corporation; Giovanni Agusta S.p.A.; Hiller Aviation Division; Hughes Helicopters; Messerschmitt-Boelkow-Blohm G.m.b.H; Robinson Helicopter Company; Sikorsky Aircraft Division. For addresses see individual listings of each company.

Airman's Information Manual—(AIM)—The operational manual for pilots containing information needed for the planning and conduct of flights in the contiguous U.S. It consists of a basic flight manual, ATC procedures, airport facilities directory, flight operation information and NOTAMs. It is published periodically by the FAA, but, because of its high price, is subscribed to by a relatively small percentage of pilots. Copies are often available to FBOs for use by pilots.

Air mass—A large body air with uniform properties of temperature and moisture content in a horizontal plain.

Air mass thunderstorm—Individual thunderstorms, widely separated from other such storms, most often found in mountainous terrain and rarely associated with frontal activity.

AIRMET—See *advisory for light aircraft.*

Air navigation facility—A facility, electronic or visual, that assists the pilot or navigator in determining his position relative to a fixed point or path on the ground.

Airplane—Any heavier-than-air craft supported by airflow over the air foils. Primarily a powered fixed-wing aircraft.

Airplane tow—In soaring, the term used for towing a glider off the ground and into the air to any predetermined altitude, using a powered aircraft. It is the most popular means of glider launch in the U.S.

Air pocket— A popular but incorrect term for descending currents of air which cause the aircraft to momentarily sink rapidly. The correct term is *downdraft.*

Airport—Any place on land or water designed to be regularly used by aircraft for takeoffs and landings; including associated buildings and facilities.

Airport Development and Assistance Fund—(ADAP)—A fund authorized by Congress and administered by the FAA for the purpose of providing financial assistance for the improvement of existing airports and the development of new airports. The funds are available on a matching-fund basis, meaning that states, counties or municipalities must come up with a given percentage of the total amount needed for the project in order to be eligible for the ADAP funds. These funds are earmarked for work actually affecting the establishment, development and improvement of the airport as such, and may not be used for terminal buildings or customer-convenience items. While originally designated to be available only to publicly owned airports, there is strong sentiment to use them also to save and improve privately owned public-use airports.

Airport advisory area—The area within five statute miles of an uncontrolled airport on which FSS is located. Within this area a pilot on approach should contact the FSS for advisories; or a pilot about to depart should inform the FSS of his intentions. Communication with such FSS is advisable but not mandatory, and its advisories should not be misunderstood as air traffic control clearances, unless prefaced by the term: *ATC clears.*

Airport advisory service—The service provided by FSS personnel operating on an airport without operating control tower for the purpose of advising arriving and departing aircraft on wind direction and velocity, active runway, altimeter setting, traffic, field conditions, standard approach procedures such as left or right traffic patterns and other facts of importance to the pilot. The service is not authorized to give takeoff or landing clearances. Frequency: 123.6.

Airport directory—A section of the AIM listing all airports available for civil-aviation use, by states. Another and more convenient listing of all airports, helipads and seaplane bases is published annually by the AOPA.

Airport elevation—(field elevation)—The listed airport elevation is the highest point of any of the useable runways, given in feet msl.

Airport identification beacon—(rotating beacon)—A beamed, green and white light timed to rotate at six rpm and usually located at or near the control tower. Beacons are operating during hours of darkness and usually (but not always) during hours of daylight when the airport is reporting weather conditions below VFR minimums. Alternating one green and one white light indicates a civil airport. Alternating one green and two white lights indicates a military or combined military-civilian airport.

Airport information desk—(AID)—An unmanned facility located at an airport and designed to facilitate self-briefing by pilots in flight preparation, planning and the filing of flight plans. AIDs are kept supplied with the latest NOTAMs, weather reports, forecasts, flight-plan forms, etc.

Airport lighting—See *runway lights.*

Airport surface detection equipment—(ASDE)—Radar which is designed especially to "read" all principal features on the airport surface, including the movements of aircraft and vehicles, presenting the resulting image on a radar display in the control tower for use by ground controllers. Usually found at large and busy airports where part of the airport cannot be seen from the tower, or where, under conditions of restricted visibility, outlying areas cannot be observed by the naked eye.

Airport surveillance radar—(ASR)—A relatively short-range radar used to control traffic in the vicinity of an airport. It provides azimuth and distance information but no altitude information. It can be used for a ground-controlled radar approach (GCA) under appropriate ceiling and visibility conditions. Such an approach is a non-precision approach as it includes no vertical guidance.

Airport traffic area—The area within a radius of five statute miles around an airport with an operating control tower, reaching upward from the ground to (but not including) 3,000 feet agl. Within an airport traffic area all VFR aircraft, whether intending to land or just passing through, must maintain two-way radio contact with the control tower. (Airport traffic areas do not affect IFR traffic). Airport traffic areas are not shown on aeronautical charts.

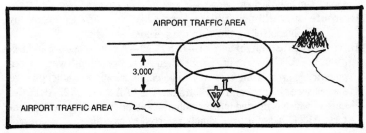

Airport traffic area (courtesy of FAA).

Airport traffic control tower cab.

Airport traffic control tower—The central operations facility serving ATC in the airport area. It consists of the tower structure and the tower cab, an IFR room if radar is used (a separate facility in the New York area where the Common IFR Room serves Kennedy International, La Guardia and Newark), and accommodates all equipment needed to signal to or communicate with aircraft. Activities performed by the tower are ground control, tower control, approach and departure control or any portion of those. Usually simply referred to as the *tower*.

Air traffic pattern—A pattern of flight around the active runway, established to provide safe flow of traffic. Patterns usually consist of a downwind leg, base leg and the final approach, in most instances involving left-hand turns all the way. At some airports right-hand patterns are in effect for some or all runways. At uncontrolled airports the segmented circle provides the pilot with the appropriate information. At controlled airports the tower will often clear pilots for abbreviated base-leg or straight-in approaches.

Air route surveillance radar—Long-range radar used primarily by ARTCCs in controlling the en-route portions of flight.

Air route traffic control center—(ARTCC)—A central operations facility in the ATC system in charge of en-route operations over a given geographic area, using two-way radio communications and long-range surveillance radar to control IFR traffic. The FAA operates 27 ARTCCs, each of which is divided into a number of sectors with discrete frequencies. ARTCCs are usually simply referred to as *centers*.

Airport traffic control tower.

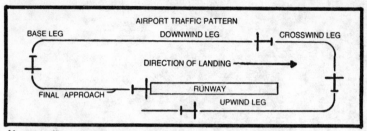

Airport traffic pattern (courtesy of FAA).

Airspace—A term usually used to mean the navigable airspace, for all practical purposes between ground level and 60,000 feet. In the United States the airspace is divided into uncontrolled airspace (airspace in which IFR operations may be conducted without contacting ATC and over which ATC has no jurisdiction), and various types of controlled airspace, affecting all IFR and in some instances VFR operations. For detailed descriptions of the various types of controlled airspace and its effect on aircraft operating in that space, see: *Airport traffic area; airport control zone; airways; control areas; control zones; continental control area; terminal control area (TCA); transition area; positive control area; uncontrolled airspace.*

Airspeed—The speed with which an aircraft is moving with relation to the air around it. Airspeed may be measured and expressed in a variety of ways: Indicated airspeed (IAS) is the speed shown in the cockpit by the

Air Route Traffic Control Center (ARTCC).

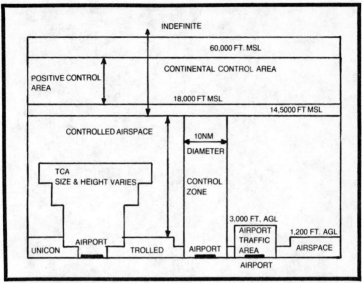

Airspace composition.

airspeed indicator. Calibrated airspeed (CAS) is indicated airspeed corrected for installation and instrument error. True airspeed (TAS) is the calibrated airspeed corrected for air density and temperature differences. It is identical to IAS at sea level under standard atmospheric conditions. At any altitude above sea level the TAS can be determined by using a standard aviation computer.

Airspeed indicator—A flight instrument with a cockpit readout which, in terms of knots or mph, shows the difference between pitot pressure and static pressure. On modern aircraft, airspeed indicators are equipped with colored circular bands: The white band indicates the speed range during which flaps may be operated, with the lower end of the arc coinciding with the stalling speed with power off and flaps and gear in landing position. The upper end of the arc represents the maximum speed with the flaps extended. The green band represents the normal operating speed range from stall speed up to the maximum structural cruising speed. The yellow arc or band is the so-called caution range, covering speeds which should be avoided in turbulent conditions. The red line indicates the never-exceed speed. In addition, on twin-engine aircraft a blue line indicates the best single-engine rate-of-climb speed and a second the red line (in the low speed area) represents minimum single-engine control speed (V_{mc}).

Airstart—Restarting an engine in flight.

Air traffic control—(ATC)—The system operated by the FAA (and to a limited extent by the military) for the purpose of promoting safe, orderly,

27

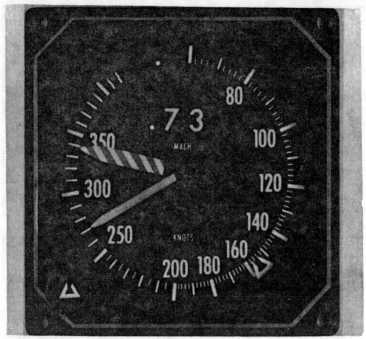

Airspeed indicator (courtesy of Smiths Industries).

and expeditious flow of air traffic. The primary obligation of ATC is to assure safe separation between IFR aircraft in controlled airspace.

Air traffic control specialist—(controller)—An employee of the FAA working in control towers, ARTCCs or related ATC facilities. (Those manning FSSs are referred to as flight service specialists). Controllers earn a salary of between $25,000 and $27,000. They must be less than 31 years old to be eligible for employment. The mandatory retirement age is 55.

Air transport rating—A pilot's license required of pilots operating airline jets in the position of captain. Also known as *ATR* or *ATP*.

Airway—Any path within the navigable airspace designated by the FAA as usable for air traffic. Usually airways are associated with ground-based nav aids. The airways most used by general aviation are the Victor airways which extend vertically from 1,200 feet agl up to but not including 18,000 feet msl. Generally they are eight miles wide and are shown on all types of aviation charts. Airways are controlled airspace.

Airworthiness directive—(AD Notes)—Notices put out by the FAA when a certain defect has been discovered during the operation of a certificated aircraft. AD Notes spell out corrective measures and deadlines within which these measure must be accomplished in order to keep the Airworthiness Certificate in effect. Compliance is mandatory.

AIS—aeronautical information service.

Alcor Aviation, Inc.—Manufacturer of EGT and engine analyzer systems. (P.O. Box 32516, San Antonio, TX 78284).

Alert area—Airspace which may accommodate a high degree of (military) pilot training activity or other types of unusual aerial activity. These areas are shown on aeronautical charts to warn non-participating pilots to be extra alert. Activities in these areas are conducted in accordance with the FARs and participating aircraft as well as those not participating, are responsible for collision avoidance.

Alert notice—(ALNOT)—A message distributed by flight service stations requesting an extensive communication search for an overdue aircraft, unreported aircraft or one that appears to be missing.

ALG—Along.

Allison—A family of turbine engines manufactured by Detroit Diesel Allison Divison of General Motors.

All weather low altitude training route—(Olive Branch Routes) (OB Routes) (AWLARs) Routes used for the training of Air Force and Navy jet pilots in IFR or VFR weather conditions, from the surface to a published altitude. Routes, altitudes and graphic representations are included in AIM Part 4. When not available, they and their current operational status may be obtained by contacting a nearby Flight Service Station.

ALNOT—Alert notice.

Alouette—A seven-place single-turbine helicopter manufactured by Aerospatiale Helicopters in France.

Aloutte II (courtesy Aerospatiale).

Alphanumerics.

ALPA—Air Line Pilots Association.

ALPHA—In aviation-radio phraseology the term for A.

Alpha-numerics—Data consisting of a combination of letters and figures.

Alpha-numeric display—The data blocks representing aircraft in flight as appearing on the controller's radar scope. These displays are generated by computers or may be manually entered by controllers. They move about the radar scope in physical relation to the actual blip.

ALS—Approach-light system.

ALT—Altitude.

ALTA—Association of Local Transport Air Lines.

Alternate airport—An airport chosen by a pilot, filing IFR, as a secondary choice, in the event that the primary destination is below landing minimums at arrival time.

Alternator—An electrical device (in recent years having replaced the old-style generator in most aircraft) which is driven by the engine and supplies current to the battery and to all electrical equipment except the ignition system. When a battery is completely drained, an alternator will not recharge it, even if the aircraft is started by propping it. In that case the battery must first receive a charge from another source.

Alternators and/or generators are manufactured and marketed by: AiResearch Manufacturing Company of California; Bendix Electric and Fluid Power Division; General Electric Aerospace; Lear-Siegler Power

Equipment Division; Simmonds Precision Products; Skytronics; Ward Aero; Westinghouse.

Altimeter—A flight instrument capable of reading height above sea level (or any other predetermined level), activated by an aneroid barometer measuring atmospheric pressure at the given altitude. This information is transformed into a readout showing feet above sea level (or any level) on a circular dial. The instrument is adjusted to the prevailing barometric pressure by turning a knob on the face of the instrument. It is a part of the pitot-static system.

Altimeter—In aviation radio phraseology the term used for altimeter setting, i.e. the prevailing barometric pressure.

Altimeter setting—The barometric pressure reading in the small window provided for that purpose on the face of the instrument, as set by the pilot in accordance with information received from a station located in the vicinity of where the aircraft happens to be located (on the ground or in the air) at that particular time.

Altitude—The height above sea level, ground level or any other reference plane. The various types of altitude of concern to the pilot are: *Indicated*

Altimeter (courtesy of Smiths Industries).

altitude—the reading shown by the altimeter when the proper altimeter setting has been made; *Pressure altitude*—the altitude above sea level (or any other given level) when the altimeter is set to 29.92; *Density altitude* is the pressure altitude corrected for prevailing temperature conditions; *True altitude*—actual height above mean sea level; *Absolute altitude*— actual altitude above the terrain over which the aircraft is flying. Of the five, those of primary interest to pilots in the every-day operation of the aircraft are: indicated altitude, which he will habitually use when flying below 18,000 feet; pressure altitude is used for all operations above 18,000 feet; and density altitude which is of primary importance when planning a departure from a high-elevation airport on a warm day. The higher the density altitude, the less efficient is the (non-turbocharged) engine, and at the same time the lifting capability of the airfoil is reduced, resulting often in very long takeoff runs or, in extreme cases, the inability of a heavily loaded aircraft to accomplish liftoff and climbout.

Altitude chamber — Pressurized facilities usually operated by the military, in which pilots may (and are encouraged to) experience the effects of oxygen starvation under high-altitude conditions without ever leaving the ground. These exercises are performed under strict supervision. They are quite safe and extremely informative. Pilots interested in participating should contact the nearest Air Force establishment.

Altitude encoder—An instrument which either optically or mechanically reads the current altimeter reading and gives this information to the transponder which, in turn, transmits it to ATC. Also referred to as the encoding altimeters or digitizers. These instruments are manufactured by: Aero Mechanism; Aerosonic Corp.; Aircraft Instrument and Development; ARC Division of Cessna; Bendix Instruments and Life Support Division; Edo-Aire Fairfield Division; HT Instruments; Instruments and Flight Research; Intercontinental Dynamics Corporation; King Radio Corp; Kollsman Instrument Co.; Narco Avionics; Smith Industries, Inc.; United Instruments.

Altitude reservation—(ALTRV)—A specialized use of the airspace, usually for the mass movement of aircraft, which would be difficult to accomplish otherwise. They must be applied for and are approved by the appropriate FAA facility.

Altitude restriction—An altitude or series of altitudes issued in order flown, and which are to be maintained until reaching a specific checkpoint or time. They are usually issued as part of ATC clearances as the result of traffic, terrain or airspace considerations.

ALTN—Alternate.

Altocumulus—A middle cumuliform cloud which forms in a layer or group of small, broken puffs. It may merge with altostratus; distinguished from cirrocumulus by the slightly grey color.

Altostratus—A middle stratiform cloud, usually in a thick, solid, greyish layer through which the sun may be vaguely seen.

Altitude encoders comparison chart.

The chart below compares encoding altimeters/digitizers. The full column structure is: Manufacturer, Model, Price uninstalled, Volt input, Display (Digitizer only / Combined altimeter-digitizer / 3-pointer / drum pointer / counter d.p. / counter point. / pointer dial / digital), Digitizer (solid state / optical / contact / magnetic), Type altim. (pneumatic / servoed), Altitude Range, Weight. (The filled check-cells for Display, Digitizer and Type are indicated by black squares in the original and are not reliably transcribable as text.)

Section	Manufacturer	Model	Price uninstalled	Volt input	Altitude Range	Weight
ENCODING ALTIMETERS/DIGITIZERS	AERO MECHANISM	AM 150	625 to 675	14 / 28	−1,000−+30,740	1.5
	AIRCRAFT INSTR. & DEVELOPMENT	371005	650 to 695	14 / 28	−1,000−+20,000 / −1,000−+30,000	2
	AEROSONIC	1019	650 to 750	14 / 28	−1,000−+20,000 / −1,000−+35,000	1.8
	AEROSONIC	1016	695 to 800	14 / 28	−1,000−+20,000 / −1,000−+35,000	1.9
	AERO MECHANISM	8140B	730 to 887	14 / 28	−1,000−+20,000 / −1,000−+35,000	1.4
	IFR	E41	795	14 / 28	−1000−−+34,000	1.12
	KOLLSMAN	ALTI-CODER II	795	14 / 28	−1,000−+35,000	2.5
	KING	KE 127	805	14 / 28	−1,000−+20,000	1
	NARCO	AR 500	810	14 / 28	−1,000−+25,000	1
	AEROSONIC	1014	900 to 1070	14 / 20	−1,000−+20,000 / −1,000−+35,000 / −1,000−+50,000	2.1
	AERO MECHANISM	8142B	995 to 1,045	14	−1,000−+35,000	2
	AERO MECHANISM	AM 165	998 to 1.169	14 / 28*	−1000−−+30,740	2.5
	KING	KEA 128	1,045	14 / 28	−1,000−+30,000	1.8
	UNITED INSTRUMENTS	5035P2	1,092.50	5 / 12 / 28	−1,000−+20,000	1.8
	SMITHS INDUSTRIES	01/200/103	1,095	14 / 28	−1,000−+20,000	1.8
	KING	KEA 125	1,150	14 / 28	−1,000−+20,000	1.9
	BENDIX INSTR. & LIFE SUPPORT	325201	1,175	14 / 28	−1,000−+35,000	2.2
	UNITED INSTRUMENTS	5035P	1,219	5 / 12 / 28	−1,000−+35,000	1.8
	KING	KEA 126	1,270	14 / 28	−1,000−+35,000	1.9
	AEROSONIC	1014	1,350 to 1,500	14 / 28	−1,000−+20,000 / −1,000−+35,000 / −1,000−+50,000	2.4
	ARC	EA 400	1,695	28	−1,000−+35,000	2.5
	EDO-AIRE/ FAIRFIELD	14 306	1,750	14 / 28	−1,000−+35,000	2.2
	AERO MECHANISM	8141B	1,780 to 1,832	14 / 28	−1,000−+35,000	2
	SMITHS INDUSTRIES	01/200/102	1,919	14 / 28	−1,000−+35,000	1.8
	IDC	571-25005	1,940	14 / 28	−1,000−+20,000	2.25
	BENDIX INSTR. & LIFE SUPPORT	3252028	2,500	14 / 28	−1,000−+35,000	2.2
	KOLLSMAN	ALTI-CODER III	2,999	28DC / 115AC / 400HZ	−1,000−+50.000	2.5
	IDC	570-24929	3,636	14 / 28	−1,000−+50,000	2.85
	ARC	EA 800	3,995	28	−1,000−+35,000	2.5
	SMITHS INDUSTRIES	1350 AM	4,142	28	−1,000−+50,000	3.25
	IDC	519-28702	5,732	28	−1,000−+35,000	3.25
	SMITHS INDUSTRIES	1403 AM	6,308	28	−1,000−+35,000	3.75
	IDC	523-29702	6,616	28	−1,000−+35,000	3.8
	IDC	518-28007	6,804	26AC / 400HZ	−1,000−+50,000	3.5
	SMITHS INDUSTRIES	1101 AM	7,650	26AC / 400HZ	−1,000−+50,000	4.1
	IDC	521-29007	7,940	26AC / 400HZ	−1,000−+50.000	3.8

counter d.p. = counter drum pointer

counter point. = counter pointer

ALTRV—Altitude reservation.

Aluminum overcast—A term humorously describing the so-called crowded-sky situation.

ALWOS—Automated low-cost weather observation station.

Ambiguity meter—The TO/FROM indicator in an OBI.

AMDT—Amendment.

AME—Aviation medical examiner.

Amecom Division, Litton Industries—Manufacturers of radio altimeters and other sophisticated avionics. (5115 Calvert Road, College Park, MD 20740).

American Jet Industries—Developers of the Hustler aircraft and parent company of Gulfstream-American Corporation. (7701 Woodley Avenue, Van Nuys, Ca 91406).

AMFI—Aviation Maintenance Foundation, Inc.

Ammeter—A readout on the instrument panel which indicates the rate at which the battery is being charged by the generator or alternator, in terms of amperes (amps).

Amp.—Ampere.

Ampere—A unit used in measuring electrical current, in terms of its strength.

AMS—Air Mass.

AMT—Amount.

Anabatic wind—Lateral air movement caused by rising air currents.

Analog computer—A computer that operates with numbers represented by directly measurable quantities, such as voltages or revolutions per minute. Recently most analog computer applications in aircraft have been replaced by digital computers because of their smaller size and weight and vastly greater capacity.

Anderson-Greenwood Aviation Corporation—Developers of the Aries T-250, a high-performance single-engine aircraft. (P.O. Box 1097, Bellaire, TX 77401).

Anemometer—An instrument for measuring wind speed, frequently of the rotating-cups type, which has cup-shaped scoops on the shaft. The cups are pushed by the wind and rotate around a shaft at a rate which varies with the speed of the wind.

Aneroid barometer—A device to measure atmospheric pressure by measuring the amount of expansion or contraction of the walls of a sealed hollow disk or diaphragm from which most of the air has been removed. It is the basic mechanism of the standard altimeter.

Angle of attack—The angle at which the chord line of the wing meets the relative wind. It determines the amount of lift developed at any airspeed.

Angle of attack indicator—A cockpit instrument which senses the angle of attack of the aircraft and displays it in the form of a cockpit readout. (Angle of attack indicators are manufactured by Conrac Corporation, Humphrey, Inc., Safe Flight Instrument Corporation, Teledyne Avionics).

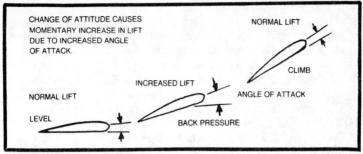

CHANGE OF ATTITUDE CAUSES MOMENTARY INCREASE IN LIFT DUE TO INCREASED ANGLE OF ATTACK.

NORMAL LIFT

CLIMB

INCREASED LIFT

ANGLE OF ATTACK

NORMAL LIFT

LEVEL

BACK PRESSURE

Angle of attack (courtesy of FAA).

Angle of bank—The angle between the lateral axis of the aircraft and the plane of the earth's surface. Also called, angle of roll.

Angle of climb—The angle between the flight path of a climbing aircraft and the horizontal plane.

Angle of glide—The vertical angle between the flight path of a descending aircraft and the horizontal plane.

Angle of incidence—The angle formed by the longitudinal axis of an aircraft and the chord line of the wing. It is a built-in, usually slightly upward angle of the wing. Also used in reference to helicopter rotor blades.

Angle of pitch—The angle between the longitudinal axis of an aircraft and the horizontal plane.

Angle of roll—Angle of bank.

Angle of yaw—The acute angle between the longitudinal axis of an aircraft and its flight path as seen from above, such as during a slip.

ANLYS—Analysis.

Annual inspection—The complete examination of an aircraft and its systems by a licensed mechanic or repair station. It is required once a year of all aircraft to maintain the airworthiness certificate in force.

ANRA—Air navigation radio aids.

ANT—Antenna.

Antenna—A device designed to send or receive radio energy. It's type and shape depend on the type of equipment to be operated in conjunction with a given antenna, and whether airborne or ground-based. (Aircraft antennas are manufactured by most major avionics manufacturers and by Aeronautical Instrument and Radio; Antenna Specialists; Comant Industries; Dayton Aircraft Products; Dorne and Margolin; Lear Siegler Instrument; Sensor Systems).

Antenna Specialists—Manufacturers of the aircraft antennas. (12435 Euclid Avenue, Cleveland, OH 44106).

Anticollision device—Collision-avoidance system.

Anticollision light—A rotating red light on an aircraft, required for night and IFR flight. Certain strobe lights may be used in conjunction with or instead of the red rotating light.

Anticyclone—A high-pressure area.

Anti-icing devices—Devices such as heated wings, pulsating leading-edge boots and alcohol systems designed to prevent ice from forming on the propeller, wings and tail surfaces of an aircraft.

Anti-torque rotor—Tail rotor on a helicopter.

Anvil cloud—A cumulonimbus cloud with an anvil-like shape at the top; typical of active and dissipating thunderstorms.

AOCI—Airport Operators Council International.

AOE—Airport of entry.

AOPA—Aircraft Owners and Pilots Association.

AP—Small hail (in sequence reports).

APCH—Approach.

Apex—The top of an inflated balloon.

Apex rope—The rope attached to the top of a balloon which, when pulled, causes a release of the hot air or gas.

API—American Petroleum Institute.

APP CON—Approach control.

Appendix—The sleeve at the bottom of a gas balloon through which gas is fed into the balloon.

Applebay Sailplanes—Developers and manufacturers of the Zuni high-performance sailplane. (206 Industrial Road, Rio Rancho, NM 87124).

Approach clearance—ATC authorization for a pilot to commence an instrument approach. The clearance includes the type of approach to be executed, plus other pertinent information.

Approach control—The ATC facility monitoring and directing traffic approaching an airport where such a facility is in operation.

Approach fix—A fix usually associated with electronic navigation aids, marking points along the final approach.

Approach gate—A position one mile from the approach fix and usually five miles from the runway threshold. Aircraft are vectored to this point to put them into position for the final touchdown.

Approach light system—Lighting which may be installed at an airport to help guide arriving aircraft. (For details see *Runway-light systems)*.

Approach path—A flight course in the vicinity of an airport, designed to bring aircraft in to safe landings. Usually, approach paths are marked by appropriate navigation aids.

Approach plate—The chart depicting an instrument approach.

Approach procedures—Published procedures describing the manner in which a precision or non-precision instrument approach must be flown.

Approach speed—The speed at which an approach to a landing is made at a relatively slow rate of descent. It is generally approximately 50 percent above the stalling speed of a given aircraft.

Apron—That part of an airport designed for tiedown, fueling, loading, etc.

APRXLY—Approximately.

ARC, Division of Cessna Aircraft Company—Manufacturers of a wide variety of avionics equipment. (Boonton, NJ 07005).

Piper Archer (courtesy of Piper).

Archer—A type of single-engine aircraft produced by Piper Aircraft Corporation.

Arctic air mass—An extremely cold air mass originating in the arctic regions.

Area broadcast—A weather broadcast made by FSSs at 15 minutes past the hour, reporting on weather conditions for an area covering approximately 150 miles from the FSS in all directions.

Area forecast—A weather forecast for a 12-hour period covering weather conditions to be expected within a region of the U.S. For this purpose the U.S. is divided into 23 regions. The area forecast is issued every six hours.

Area navigation—A method of navigating without reference to the established airways. This method involves the need for a variety of on-board avionics instrumentation. The most popular method, useable only within reception distance of the North American VORTAC network, requires one (or several) VOR receivers, a DME and a CLC (course-line-computer). With these instruments the pilot can electronically relocate VORs, placing them on any desired location within approximately 60 miles of their actual location. The other two area-navigation systems involve long-range navigation equipment and can be used anywhere in the world. They are the *inertial guidance system* (INS) and *VLF/Omega*, and are described under those headings.

Ariel—A turbine engine manufactured by Turbomeca in France.

Aries T-250—A high-performance single-engine aircraft developed by Anderson-Greenwood Aviation Corporation and marketed by Bellanca Aircraft Corporation.

Area navigation equipment comparison chart.

MANUFACTURER	MODEL	PRICE estimated	Volts input	INTERFACE VOR	INTERFACE DME	DISPLAY	WAYPOINTS Number	WAYPOINTS Range nm	ENTRY	Weight	Remarks
FOSTER AIRDATA	RNAV 511	1.649	14 28	ALL	ALL		2	199.9		2.5	
KING	KN 74	2.503	14 28	KING	KING		1	199		4	OUTPUTS FOR AUTOPILOT, FLIGHT DIRECTOR, RMI
KING	KNC G10	2.995	14 28	MOST	MOST		1	150		3.7	OUTPUTS FOR AUTOPILOT, FLIGHT DIRECTOR, RMI. VNAV OPTION
COLLINS	ANS-351	3.120	14 28	COLLINS KING	COLLINS KING		8	199		3.3	PRESENT POSITION & DIGITAL BEARING READOUT
	RN-400A	3.695	14 28	ARC 300 400	ARC 400		3	199.9			RMI TO WP OUTPUT
FOSTER AIRDATA	AD G11	4.000 to 7.000	14 28	ALL	ALL		to 11	299.9		10	PRICE VARIES W OPTIONS VNAV OPTIONAL
NARCO	RNAV 161	4.235		ALL	NARCO		10	199.9		5.6	NON-VOLATILE MEMORY OUTPUT FOR AUTOPILOT AND HSI
ARC	RN-800A	4.995	28	ARC 1000	ARC 800		5	199.9			RMI TO WP OUTPUT NON-VOLATILE MEMORY OPT.
KING	KNS 80	5.665	14 28	INCL.	INCL.		4	199.9			COMPLETE SYSTEM INCL. GLIDE SLOPE. 200-CH. VOR/LOC. NON VOLATILE MEMORY
KING	KNR 665	10.000	28DC 26AC 400HZ	INCL	KDM 705		10	150		4.4 13.8	OUTPUTS FOR AUTOPILOT. FLIGHT DIRECTOR
BENDIX AVIONICS	RNS 3500	11.034	28	MOST	MOST		32	150		8.1	
COLLINS	ANS 31	13.001	28	ALL	ALL		10	249.9		11.4	PUSHBUTTON PROGRAMMING TUNES NAV & DME
JET ELECTRONICS	DAC-Z000	13.224	28	ALL	ALL		200	999.9		15.6	VNAV INCLUDED
EDO-AIRE/ FAIRFIELD				ARINC 547 579	ARINC 521C 568		20	499.9		28	
COLLINS		NA	28		ALL			NA		14.5	TOUCH-TUNES ALL RADIOS
COLLINS	ANS-70A	NA	115AC 400HZ	ARINC	ARINC		NO LIMIT			87	WORLD-WIDE NAV-AID STRUCTURE PRE-STORED IN NON-VOLATILE MEMORY

Dist. to WP = distance to waypoint. Ground sp. = ground speed. Time to WP = time to waypoint.

ARINC—Aviation Radio, Inc., an organization which sets a variety of standards for high-performance avionics, especially for equipment designed for the airlines and the top-of-the-line corporate turbine aircraft.

ARND—Around.

ARPT—Airport.

ARR—Arrival; arrive.

Arresting system—A variety of mechanical means by which aircraft can be stopped and prevented from overrunning the end of the runway. The most prevalent is the arresting cable, a steel cable stretched accross the runway (or the deck of an aircraft carrier) designed to engage the arresting hook of military aircraft.

Arrival time—The touchdown time of an arriving aircraft.

Arrow—A high-performance four-seat single-engine aircraft, part of the Cherokee family of aircraft, produced by Piper Aircraft Corporation.

ARSR—Air-route surveillance radar.

ARTCC—Air route traffic control center. Usually simply referred to as *center*. The ATC facility handling en-route IFR traffic. There are 27 such centers in the U.S.

Articulated rotor—Helicopter main rotor with individual flapping, lead-lag and feathering hinges.

Artificial horizon—A gyro instrument showing the attitude of the aircraft with reference to pitch and roll as compared to the horizon.

Aries T-250 (courtesy of Bellanca).

ARTS I, II, III—Automated radar terminal system, a system of terminal air-traffic control based on sophisticated computer technology and radar. The I, II and III refer to the degree of automation at a given facility.

Artouste—A turbine engine manufactured by Turbomeca in France.

ARV—Air recreational vehicle. A joint AOPA/EAA project to develop a low-cost simple aircraft.

Piper Arrow (courtesy of Piper).

ASDE—Airport surface detection equipment.

ASL—Above sea level. Usually expressed msl.

Aspect ratio— The aspect ratio of an aircraft wing is defined as the square of the wing span divided by the wing area.

ASR—Airport surveillance radar.

ASRS—Aviation safety reporting system. A function of NASA.

Asymmetrical power—The situation existing in a twin-engine aircraft, when one engine is inoperative. This does not include the push-pull type twins such as the Cessna Skymaster, which have one engine located each fore and aft.

AStar—The single-engine turbine-powered helicopter produced by Aerospatiale Helicopter Division in France.

Astazou—A turbine engine manufactured by Turbomeca in France.

ATA—Air Transport Association—A powerful trade and lobbying organization representing the airline industry.

ATC—Air traffic control.

ATCA—Air Traffic Control Association.

ATCAA—ATC assigned airspace.

ATC advises—A term used to preface non-control information relayed by ATC to the pilot via a person other than an air-traffic controller.

ATC assigned airspace—Airspace defined in terms of lateral and vertical limits for the purpose of separating special-purpose activity (usually military) from other IFR traffic.

ATC clearance—An authorization by ATC with reference to the course and altitude to be flown by an IFR aircraft designed to assure separation between various IFR traffic.

ATC clearance limit—The farthest point to which an IFR aircraft has been cleared and beyond which it may not proceed without having received a further clearance.

ATC clearance shorthand—A convenient means of copying a clearance while it is being issued by ATC. It is non-mandatory and different pilots may use different types of such shorthand.

ATC clears—A term used to preface an ATC clearance when it is relayed to the pilot by a person other than an air-traffic controller.

ATC instruction—Instructions issued by ATC in order to cause the pilot to take a given action, such as: *Go around* or *Turn right to 180 degrees.*

ATCPA—Air Taxi and Commercial Pilots Association.

ATC requests—A term used to preface an ATC request when it is relayed to the pilot by a person other than an air-traffic controller.

ATCRBS—Air traffic control radar beacon system.

ATIS—Automatic terminal information service. Recorded information about weather and other conditions at the airport, periodically updated when conditions change. Each such report is prefaced by *Information alpha (bravo*, etc.) and the pilot is expected to inform tower, ground or approach control that he has received information alpha (bravo, etc.).

Words and Phrases	Shorthand
ABOVE	ABV
ADVISE	ADV
AFTER (PASSING)	<
AIRPORT	A
(ALTERNATE INSTRUCTIONS)	( )
ALTITUDE 6,000-17,000	60-170
AND	&
APPROACH	AP
FINAL	F
LOW FREQUENCY RANGE	R
OMNI	O
PRECISION	PAR
STRAIGHT-IN	SI
SURVEILLANCE	ASR
APPROACH CONTROL	APC
AT (USUALLY OMITTED)	
(ATC) ADVISES	CA
(ATC) CLEARS OR CLEARED	C
(ATC) REQUESTS	CR
BEARING	BEAR
BEFORE	>
BELOW	BLO
BOUND	B
EASTBOUND, etc.	EB
INBOUND	IB
OUTBOUND	OB
CLIMB (TO)	↑
CONTACT	CT
CONTACT DENVER	
APPROACH CONTROL	DEN
CONTACT DENVER CENTER	(DEN
COURSE	CRS
CROSS	X
CROSS CIVIL AIRWAYS	≠
CRUISE	⟶
DELAY INDEFINITE	DLI
DEPART	DEP
DESCEND (TO)	↓
DIRECT	DR
EACH	ea
EXPECT APPROACH	
CLEARANCE	EAC
EXPECT FURTHER	
CLEARANCE	EFC
FAN MARKER	FM
FLIGHT PLANNED ROUTE	FPR
FOR FURTHER CLEARANCE	FFC
FOR FURTHER HEADINGS	FFH
HEADING	HDG
HOLD (DIRECTION)	H-W
IF NOT POSSIBLE	or
INTERSECTION	XN
(ILS) LOCALIZER	L
MAINTAIN OR MAGNETIC	M
(MAINTAIN) VFR	
CONDITIONS ABOVE:	
ALL CLOUDS	VFR
ALL HAZE	VFR/H
ALL DUST	VFR/D
ALL SMOKE	VFR/K
ALL FOG	VFR/F
OMNI (RANGE)	O
OUTER COMPASS LOCATOR	LOM
OUTER MARKER	OM
OVER (Iden.)	OKC-0
RADAR VECTOR	R V
RADIAL	RAD
RANGE (LF/MF)	R
REMAIN WELL TO	
LEFT SIDE	LS
REMAIN WELL TO	
RIGHT SIDE	RS
REPORT DEPARTING	RD
REPORT LEAVING	RL
REPORT ON COURSE	R-CRS
REPORT OVER	RO
REPORT PASSING	RP
REPORT REACHING	RR
REPORT STARTING	
PROCEDURE TURN	RSPT
REQUEST ALTITUDE	
CHANGES ENROUTE	RACE
REVERSE COURSE	RC
RUNWAY	RY
STANDARD JET	
PENETRATION	SJP
STANDBY	STBY
TAKEOFF (DIRECTION)	T →N
TOWER	Z
TRAFFIC IS	TFC
TRACK	TR
TURN LEFT	
AFTER TAKEOFF	LT
TURN RIGHT	
AFTER TAKEOFF	RT
UNTIL	/
UNTIL ADVISED (BY)	UA
UNTIL FURTHER ADVISE	UFA
VICTOR	V
WHILE IN CONTROL AREAS	Δ
WHILE IN CIVIL AIRWAYS	=

EXAMPLE

ATC clears (Iden.) to St. Louis Airport via Victor 16, Victor 9. Maintain one five thousand. Depart south. Turn left after departure. Proceed direct Dallas Omni. Cross Dallas Omni at 3000. Report leaving 3, 4, and 5000.
C STL-A V16 V9. M 150. TS. LT. DR DAL-OXDAL-O 30. RL 3, 4, 50.

ATC clearance shorthand (courtesy of FAA).

Atmosphere—The total mass of air surrounding the earth in layers of varying characteristics. A unit of atmospheric pressure is equal to a column of 29.92 inches of mercury, also expressed as 1613.2 millibars.

Atmospheric pressure—See atmosphere.

Atmospherics—Static.

ATP—Air transport pilot rating.

ATR—See ATP.

Attitude—The position of an aircraft in relation to a given reference, usually the ground, along its longitudinal, lateral and vertical axes.

Audio selector panel (courtesy of Narco).

Attitude indicator—See artificial horizon.

Audio selector panel—An avionics device making it simpler for the pilot to select the particular nav or com radio he wants to use at a given time. Audio selector panels are produced by all major avionics manufacturers.

AUTO—Automatic.

Automated radar terminal systems—See ARTS.

Automatic altitude reporting—The function of a transponder, designated Mode C, which accepts altitude information from the encoding altimeter and/or digitizer, and automatically transmits it to ATC in 100-foot increments.

Audio selector panels comparison chart.

MANUFACTURER	MODEL	PRICE uninstalled	Volts input	Input transceiver	Input receiver	Marker beacon built in	Marker beacon separate	Selector switch VHF com	Selector switch HF com	Selector switch PA system	Selector switch intercom	Auto feature	Solid state	Size (inches) wide/high/deep	Weight lbs.	REMARKS
EDO-AIRE FAIRFIELD	A-550	250	14 28											6.5×1.5×7.43	1.8	
KING	KA 134	290	14 28	3	5		*							5.95×1×6.25	0.8	*MARKER BEACON $165
GENAVE	TAU 200	295	14 28	3	4									6.5×1.6×6.75	1.8	
MENTOR	AP-1	300	14 28											6.36×1.6×7.38	1.5	
COLLINS	AUD-250	315	14	3	4				OPT	OPT				6.25×1.6×5.6	1.5	
BENDIX AVIONICS	AS-2015A	340	14 28											6.26×1.86×4.21	1	
NARCO	CP-135	340	14 28	2	4	OPT			OPT					6.25×1.1×8.3	1.8	
NARCO	CP-136	435	14 28	2	4	OPT			OPT					6.25×1.1×8.3	2	
MENTOR	APM-1	450	14 28											6.36×1.6×7.38	1.6	AUTOMATIC DIMMING OF MARKER LIGHTS
COLLINS	AUD 25a H	485	14	3	4									6.25×1.6×5.6	NA	TSO FIVE STATION INTERCOM
EDO-AIRE FAIRFIELD	AM-550	525	14 28											6.5×1.5×7.34	2.1	AUTOMATIC SPEAKER MUTING
COLLINS	AMR-350	575	14	3	4									6.25×1.6×5.6	1.8	
EDO AIRE FAIRFIELD	AM-660	595	14 28											6.5×1.5×7.34	2.1	AUTOMATIC SPEAKER MUTING
NARCO	CP-200	600	14 28*	2	4									6.25×1.1×8.3	2.1	*WITH CONVERTER AUDIO ALERT FOR OPTIONAL E-LINE KEYBOARD
NARCO	CP-136T	610	14 28	2	4	OPT								6.25×1.1×8.3	2.1	
KING	KMA 20	630	14 28	2*	5	OPT		*						3.2×1.05×7.13	2.3	TSO-*INTERNATIONAL MODEL HAS HF TRANSCEIVER INPUT AT $30 EXTRA
COLLINS	346 B-3	1,990	28					?	?	?	?			REMOTE MOUNTED	3.5	TSO TOTAL INPUTS 12
BAKER	M1035	NA	28	4	11									5×2.25×5	2.2	
BAKER	M1045	NA	28	6	16									5×2.62×5.19	2.5	

Automatic direction finder with loop-sense antenna (courtesy of Collins).

Automatic direction finder—ADF. A cockpit instrument which responds to radio signals from non-directional beacons (NDB), standard broadcast stations and a variety of other ground-based LF/MF navigation aids, translating this information into a cockpit readout, the needle of

Automatic direction finders comparison chart.

MANUFACTURER	MODEL	PRICE uninstalled	Volts input	Frequency range kHz	Panel mounted	Remote mounted	Digital tuning	BFO	Size (inches) wide/high/deep	Units	Weight lbs.	REMARKS
H.T. INSTRUMENTS	ACCU TUNE	195	14 28	100-2.500					2×1.5×3	1	0.3	DIGITAL FREQUENCY INDICATOR ONLY INTERFACES WITH NON-DIGITAL ADFs
GENAVE	SIGMA/1500	995	14	200-1.699					6.1×2×8.6	2	4.5	ROTATABLE HEADING CARD
EDO-AIRE/FAIRFIELD	R566E	1,325	14 28*	188-1.750					6.5×3×8.75	2	8.5	*ADAPTER NEEDED INCLUDES ELECTRIC ELAPSED TIMER NO RMI CAPABILITY
KING	KR 86	1.465	14 28	200-1.750					6.25×2.6×9.05	1	8.9	*AVAILABLE W COMBINED LOOP SENSE ANTENNA AT NO EXTRA COST
ARC	R-300	1.595	28	200-1.699					6.6×2.9×13.25	3	5	
COLLINS	ADF 650	1.650	14	200-1.799					6.25×1.75×9	3	5.3	COMBINED LOOP SENSE ANTENNA
NARCO	ADF 141	1.690	14 28*	200-1.799					6.25×1.75×9	2	6.8	*WITH CONVERTER
KING	KR85	1.885	14 28	200-1.699					6.2×2.6×9.32	1	6.6	TSO AVAILABLE W COMBINED LOOP SENSE ANTENNA AT NO EXTRA COST
BENDIX AVIONICS	ADF 2070	1.965	14 28*	200-1.799					6.26×1.75×9.2	2	5.8	*28 V ADAPTER $15 EXTRA
ARC	R-400	2.295	28	200-1.699					6.6×2.9×13.25	3	5	DUAL FREQUENCY PRESELECT
NARCO	ADF-200	2.600	14 28*	200-1.799					6.25×1.75×9	2	6.9	*WITH CONVERTER DUAL FREQUENCY PRESELECT KEYBOARD CHANNELING AVAIL
ELECTRONIQUE AEROSPATIALE	AD 850 R	2.800	28 DC 26 AC 400 Hz	190-1.790					5.7×2.6×12.1	1	5	
ELECTRONIQUE AEROSPATIALE	AD 850 CR	3.100	28 DC 26 AC 400 Hz	190-1.750					5.7×2.6×12.1	1	5	
KING	KDF 805	3.120	28	200-1.799					5.1×3×10	3	5.8	WITH COMBINED LOOP SENSE ANTENNA WITHOUT KFS 580 B CONTROL
ELECTRONIQUE AEROSPATIALE	AD 850	3.250	28 DC 26 AC 400 Hz	190-1.750					5.7×2.6×12.1	1	5.5	FOUR-FREQUENCY STORAGE AND RECALL CAPABILITY
KING	KDF 8000	3.328	28	200-1.750					NA	3	9	ARINC CHARACTERISTIC 570
ELECTRONIQUE AEROSPATIALE	AD 850C	3.600	28 DC 26 AC 400 Hz	190-1.750					5.7×2.6×12.1	1	5.5	FOUR-FREQUENCY STORAGE AND RECALL CAPABILITY LOOP SENSE ANTENA INTEGRATED
KING	KDF 850	3.620	28	200-1.799					5.1×3×10	3	5.8	WITH COMBINED LOOP SENSE ANTENNA WITH KFS 580 B CONTROL
COLLINS	ADF 60A	3.625	28	190-1.749.5					NA	2	5.7	COMBINED LOOP SENSE ANTENNA
ARC	R-1000	4.595	28	200-1.699					6.5×3.06×9.75	4	9.5	DUAL FREQUENCY PRESELECT
COLLINS	DF 20G	7.632 TO 9.156	27.5	190-1.750					NA	2	APPROX 20	ARINC CHARACTERISTIC 570 ADF-SYSTEM RMI CAPABILITY
BENDIX AVIONICS	DFA 74	8.048	28 DC 26 AC	190-1.750					NA	5	9.8	ARINC CHARACTERISTIC 570
MARCONI AVIONICS	AD 380	8.260	28	190-1.599.5					NA	1	10	FREQUENCY SPACING ½ kHz AVAILABLE RANGE TO 2.182 kHz

Autopilot controls (courtesy of Collins).

which points to the station, thus telling the pilot the position of the nose of the aircraft with relation to the station.

Automatic pilot—See autopilot.

Automatic terminal information service—See ATIS.

Autopilot—A device, usually consisting of gyroscopic, electronic and/or hydraulic elements which operates the flight controls of an airplane. Autopilots come in a wide variety of degrees of sophistication and prices. (See chart).

Autorotation—The condition caused by engine failure in a helicopter. The gravity pull will cause the rotor blades to rotate and thus to slow the power-off descent. Autorotation can also be achieved by pulling the throttle back to idle power.

Auto tow—A means of launching a sailplane, using an automobile to tow the aircraft until it becomes airborne.

AUX—Auxiliary.

AVBL—Available.

AVG—Average.

Avian Balloon Company—A manufacturer of hot-air balloons. (South 4323 Locust Road, Spokane, WA 99206).

Aviation medical examiner—A licensed physician designated by the FAA to perform medical examinations and issue medical certificates.

Aviation organizations—Listed here are the more prominent aviation organizations. In addition, there are literally hundreds of minor organizations, some national, regional or local.

Aero Club of Washington, 1629 K Street NW, Washington, DC 20006.

Aerobatic Club of America, P.O. Box 401, Roanoke, TX 76262.

Aeronautical Radio, Inc. (ARINC), 2551 Riva Road, Annapolis, MD 21401.

Aerospace Education Foundation, 1750 Pennsylvania Avenue NW, Washington D.C. 20006.

Aerospace Industries Association (AIA), 1725 DeSales Street NW, Washington, D.C. 20006.

Air Force Association, 1750 Pennsylvania Avenue Washington, D.C. 20006.

Air Taxi and Commercial Pilots Association (ATCPA), P.O. Box 4412, Washington, D.C. 20006.

Air Traffic Control Association (ATCA), 525 School Street SW, Washington, D.C. 20006.

Air Transport Association of America (ATA), 1709 New York Avenue NW, Washington, D.C. 20006.

Autopilots comparison chart.

MANUFACTURER	MODEL	PRICE uninstalled	Volt input	Weight lbs.	REMARKS
BRITTAIN	LEVEL MATIC	895	14/28	10	*ON SOME AIRCRAFT
BRITTAIN	ACCU TRAK II	1,095	14/28	9	*ON SOME AIRCRAFT
EDO-AIRE MITCHELL	CENTURY I	1,247	14/28	7	
ASTRONAUTICS	P-1	1,400	14/28	10.6	
ARC	AF 200A	1,515	28	9.7	
EDO-AIRE MITCHELL	YAW DAMPER	1,693	14/28	5	
BRITTAIN	ACCU FLITE II	2,195	14/28	16	*ON SOME AIRCRAFT
ASTRONAUTICS	P-2	2,233	14/28	11.5	
EDO-AIRE MITCHELL	CENTURY II	2,543	14/28	9	SLAVED GYRO, HSI, RADIO COUPLERS OPTIONAL
BRITTAIN	NAV FLITE II	2,995	14/28	20	*ON SOME AIRCRAFT
ASTRONAUTICS	P-2A	3,008	14/28	15.2	
ARC	AF 300A	3,090	28	12	*ON SOME AIRCRAFT
BRITTAIN	B-5	4,795	14/28	25	$1,000 EXTRA FOR MULTI-ENGINE
ARC	AF-400	5,550	28	19.1	SLAVED GYRO OPTIONAL
ASTRONAUTICS	P-3	5,790	14/28	23.5	
EDO-AIRE/MITCHELL	CENTURY III	6,079	14/28	18	SLAVED HSI, YAW DAMPER, GLIDESLOPE AND RADIO COUPLER OPTIONAL
ASTRONAUTICS	P-3A	6,952	14/28	26.5	
EDO-AIRE/MITCHELL	CENTURY IV	8,739	14/28	25	LESS HEADING SYSTEM ARINC 360A OPTIONAL
ASTRONAUTICS	P-3B	8,765	14/28	26.9	
ARC	AF-400A	9,150	28	23.4	HSI OR SLAVED D.G. OPTIONAL
ARC	AF-400B	9,605	28	25.5	HSI OR SLAVED D.G. OPTIONAL YAW DAMPER OPT.
COLLINS	AP-106A	14,627	28	27.7	YAW DAMPER $1,824 EXTRA GS, RNAV, BC, HSI, FD COUPLER
COLLINS	AP-105	15,275	28 / 115 AC 400 Hz	34	YAW DAMPER OPT. COUPLER TO: GS RNAV, INS, BC, HSI, FD. CONTROL WHEEL SYNC. MACH HOLD OPT.
COLLINS	APS-80	NA	28 / 115 AC 400 Hz	27.7	YAW DAMPER OPTIONAL. INTERFACES WITH FDS-84, FDS-108, FDS-109 FLIGHT DIRECTOR SYSTEMS

Column groupings in the chart (indicated by filled cells): Gyros (directional, attitude, rate); Surface control (ailerons, elevator, rudder); Control (lateral, pitch); Command turns; Hdg. (hold, preset); Alt. (hold, preset); Radio coupl (Crosswind corr., VOR, LOC); Servo (electric, pneumatic).

Aviation organizations

Aircraft Electronics Association (AEA), 6310 General Twining Avenue, Sarasota, FL 33580.

Aircraft Owners and Pilots Association (AOPA, 7315 Wisconsin Avenue, Washington, D.C. 20014.

Airport Operators Council International, Inc. (AOCI), 1700 K Street NW, Washington, D.C. 20006.

American Association of Airport Executives (AAAE), 2029 K Street NW, Washington, D.C. 20006.

American Bonanza Society, Chemung County Airport, Horseheads, NY 14845.

American Institute of Aeronautics and Astronautics (AIAA), 1290 Avenue of the Americas, New York, NY 10019.

American Meteorological Society, 45 Beacon Street, Boston, MA 02108.

American Navion Society, Municipal Airport, Banning, CA 92220.

American Petroleum Institute (API), 2101 L Street NW, Washington, D.C. 20037.

Antique Airplane Association (AAA), P.O. Box H, Ottumwa, IA 52501.

Association of Aviation Psychologists (AAP), Naval Safety Center, Code 1157, NAS Norfolk, VA 23511.

Association of Balloon and Airship Constructors (ABAC), 3217 North Delta Avenue, Rosemead, CA 91770.

Aviation Distributors and Manufacturers Association (ADMA), 1900 Arch Street, Philadelphia, PA 19103.

Aviation Hall of Fame, Dayton Convention and Exhibition Center, Dayton, OH 45402.

Aviation Maintenance Foundation, Inc. (AMFI), P.O. Box 739, Basin, WY 82410.

Aviation/Space Writers Association (AWA), Cliffwood Road, Chester, NJ 07903. c/o Wm. F. Kaiser.

Civil Air Patrol, Maxwell AFB, AL 36112.

Experimental Aircraft Association (EAA), P.O. Box 229, Hales Corners, WI 53130.

Flight Safety Foundation, Inc., 1800 North Kent Street, Arlington, VA 22209.

Flying Architects Association, c/o H.B. Southern, 571 East Hazelwood Avenue, Rahway, NJ 07065.

Flying Chiropracters Association, 215 Belmont Street, Johnstown, PA 15904.

Flying Dentists Association, c/o Jean Replogle, 5410 Wilshire Boulevard, Los Angeles, CA 90036.

Flying Funeral Directors of America, 811 Grant Street, Akron, OH 44311.

Flying Physicians Association, 801 Green Bay Road, Lake Bluff, IL 60044.

General Aviation Manufacturers Association (GAMA), 1025 Connecticut Avenue NW, Washington, D.C. 20036.

Helicopter Association of America (HAA), 1156 15th Street NW, Washington, D.C. 20005.

Avions Marcel Dassault-Breguet jet aircraft specifications.

AVIONS MARCEL DASSAULT-BREGUET MODEL:	FALCON 10	FALCON 20F	FALCON 50
ENGINES:	GA/TFE 731-2 (2)	GE/CF700-2D-2(2)	GA/TFE 731-3 (3)
THRUST (pounds each)	3,230	4,500	3,700
SEATS crew + passengers	2 + 6 to 8	2 + 8 to 10	2 + 10 to 12
BALANCED FIELD LENGTH, ISA (ft)	4,470	5,100	4,900
ISA + 20 (ft)	6,800	7,180	6,000
RATE OF CLIMB (fpm)	4,450 (1 eng. 1,050)	3,650 (1 eng. 900)	3,526 (2 eng. NA)
V_r (knots)	118	124	NA
V_2 (knots)	123	125	NA
SERVICE CEILING (ft)	45,000 (1 eng. 18,000)	42,000 (1 eng. 15,000)	45,000 (3 eng. 28,000)
Mmo	.87	.88	.85
ALTITUDE CHANGEOVER (FL)	275	234	234
Vmo (knots)	350	350	370
CRUISE hi speed (knots)	462 (FL 390)	432 (FL 370)	465 (FL 370)
long range (knots)	421 (FL 390)	395 (FL370)	425 (FL 370)
RANGE, long range cruise, w.res. (nm)	1,900 (FL 350)	1,815 (FL 350)	3,650 (FL 350)
FUEL FLOW per engine, hi speed (pph)	635	990	763
long range (pph)	545	825	590
RAMP WEIGHT (lbs)	18,740	28,660	37,480
TAKEOFF/LANDING WEIGHT (lbs)	18,740/17,640	28,660/27,320	37,480/34,850
ZERO FUEL WEIGHT (lbs)	13,560	19,600	17,640
USEFUL LOAD (lbs)	7,940	11,600	18,545
FUEL CAPACITY (lbs)	5,910	9,180	15,320
MAX PAYLOAD	2,310	1,990	2,350
FUEL w. max payload (lbs)	5,230	9 180	15,320
PAYLOAD w. full fuel (lbs)	1,630	1,990	2,350
WING LOADING (lbs per sq ft)	72.2	65.1	77.4
WING AREA (sq ft)	259.4	440	504.1
LENGTH/HEIGHT/SPAN external (ft)	45.5/15.1/42.9	56.25/17.47/53.47	60.75/22.9/61.87
LENGTH/HEIGHT/WIDTH cabin (ft)	12.7/4.9/4.7	23.3/5.7/6.2	22.1/5.9/6.1
PRESSURIZATION (psi)	8.8	8.3	8.8
EXTERIOR NOISE (EPNdB)	79.6/95.3	90/103	87/97
PRICE (1979 $s)	2,280,000	3,665,000	5,730,000

International Flying Farmers, Mid-Continent Airport, Wichita, KS 67209.

Lawyer-Pilots Association, P.O. Box 427, Alhambra, CA 91802.

Lighter-than-Air Society (LTA), 1800 Triplett Boulevard, Akron, OH 44306.

National Aeronautics Association (NAA), 806 15th Street NW, Washington, DC 20005.

National Agricultural Aviation Association, Suite 459, National Press Building, Washington, D.C. 20045.

National Air Transportation Associations, Inc. (NATA), 1156 15th Street NW, Washington D.C. 20005.

National Association of Flying Instructors (NAFI), P.O. Box 20204, Columbus, OH 43220.

National Association of Priest Pilots, 5157 South California Avenue, Chicago, IL 60632.

National Association of State Aviation Officials (NASAO), 1000 Vermont Avenue NW, Washington D.C. 20005.

National Business Aircraft Association, Inc. (NBAA), One Farragut Square South, Washington, D.C. 20005.

National Pilots Association, 805 15th Street NW, Washington, D.C. 20005.

Ninety-Nines, Inc., P.O. Box 59965, Oklahoma City, OK 73159.

Soaring Society of America (SSA), P.O. Box 66071, Los Angeles, CA 90066.

Society of Automotive Engineers (SAE), 400 Commonwealth Drive, Warrendale, PA 15096.

United Air Racing Association, 16644 Roscoe Boulevard, Van Nuys, CA 91406.

United States Parachute Association, 806 15th Street NW, Washington, D.C. 20005.

Whirly-Girls, Inc., Suite 700, 1725 DeSales Street NW, Washington, D.C. 20036.

Wings Club, Biltmore Hotel, Madison and 43rd Street, New York, NY 10017.

Aviation Publications—Listed here are only national publications. There are large numbers of regional aviation publications and newsletters.

A/C Flyer, Ziff-Davis Publishing Co., One Park Avenue, New York, NY 10016

Aero, 2377 South El Camino Real, San Clemente, CA 92672.

Air Progress, 7950 Deering Avenue, Canoga Park, CA 91304.

Airline Executive, 818 18th Street NW, Washington, D.C. 20006.

Airport Services Management, 731, Hennepin Avenue, Minneapolis, MN 55403.

AOPA Pilot, 7315 Wisconsin Avenue, Washington, D.C. 20014.

Aviation Consumer, P.O. Box 4327, Greenwich, CT 06830.

Aviation Convention News, Pan American Building, Teterboro Airport, Teterboro, NJ 07608.

Aviation Week and Space Technology, 1221 Avenue of the Americas, New York, NY 10020.

Business and Commercial Aviation, Hangar C-1, Westchester County Airport, White Plains, NY 10604.

Flying, One Park Avenue, New York, NY 10016.

General Aviation News, P.O. Box 1094, Snyder, TX 79549.

Plane and Pilot, P.O. Box 1136, Santa Monica, CA 90406.

Private Pilot, 2377 South El Camino Real, San Clemente, CA 92672.

Rotor and Wing, News Plaza, Peoria, IL 61601.

Professional Pilot, West Bldg., Nat'l Airport, Washington, DC 20001.

Soaring, P.O. Box 66071, Los Angeles, CA 90066.

Sport Aviation, P.O. Box 229, Hales Corners, WI 53130.

Trade-a-Plane, Crossville, TN 38555.

Aviation weather service—The service is provided by the National Weather Service and the FAA which collect and disseminate pertinent weather information.

Avionics—A catch-all phrase for communication, navigation and related instrumentation in an aircraft.

Avionics management systems—Highly sophisticated electronic instrumentation which collects a variety of navigation data automatically, feeds them into a computer which, in turn, controls the flight-director and autopilot systems.

Piper Aztec (courtesy of Piper).

Avions Marcel Dassault-Breguet—A major aircraft manufacturing company in France, partly government owned. Manufacturers of the line of Falcon Jet aircraft.

AW—Winds aloft forecast.

AWA—Aviation/Space Writers Association.

AX—A designation by the Federation Aeronautique Internationale (FAI) referring to various categories of balloons.

Axes—Plural of axis.

Axis—The theoretical line extending through the center of gravity of an airplane along the longitudinal, lateral and vertical planes.

Azimuth—Bearing, as measured clockwise from the true or magnetic north (except in celestial navigation).

Aztec—A six-place twin-engine piston aircraft produced by Piper Aircraft Corporation.

B—Balloon ceiling (weather reports only).

B—Beginning of precipitation; followed by time in minutes. (Weather reports only).

B—Bravo in the phonetic alphabet.

B—Ceiling as measured by a balloon (in sequence reports.)

B—Scheduled weather broadcast.

Backcourse—The reverse side of a localizer. A backcourse approach is a non-precision approach along the backcourse of a localizer.

Backcourse approach—See backcourse.

Backing—Wind shifting in a counter-clockwise direction, to the left of the direction from which it was blowing before (the opposite of veering).

Bail out—Jumping out of an aircraft with or without (preferably with) a parachute.

Balance—The stability of an airplane achieved when the four forces, thrust, drag, lift and weight (gravity) act to produce steady flight.

Balanced field length—The distance within which a jet aircraft can accelerate to V_1 and then either stop or accelerate to a safe climb speed (V_2) and clear a height of 35 feet on one engine.

Ballast—Weight, usually bags of sand, carried aloft in gas balloons as a means of maintaining altitude by jettisoning ballast.

Balloon Federation of America (BFA)—National organization of sports balloonists.

Balloon license—License issued by the FAA to properly trained balloon pilots.

Balloon Works—Manufacturers of hot-air balloons. (Rhine Aerodrome, RFD 2, Statesville, NC 28667).

Bank—To tilt an airplane by means of the ailerons, causing it to turn while rolling either right or left along its longitudinal axis.

51

Bar—A unit of pressure equal to 29.531 inches of mercury at 32 degrees F. (0 degrees C.) at 45 degrees latitude.

Barnstorming—An activity of pilots who used to fly into small communities and take passengers up for short rides for a fee. Barnstorming, prevalent in the 1920s and 30s, has virtually vanished from the aviation scene.

Barometer—An instrument for measuring atmospheric pressure. Mercurial barometers measure the effect of atmospheric pressure on a column of mercury. Aneroid barometers detect pressure changes in the partial vacuum of a hollow disc.

Barometric pressure—Atmospheric pressure measured by a barometer.

Barometric tendency—The net change in barometric pressure during the last three hours before the observation; given in station reports as + or − a given number of tenth of millibars.

Baron—A six- or eight-seat high-performance piston twin manufactured by Beech Aircraft Corporation.

Barrel roll—An aerobatic maneuver in which the aircraft, while maintaining its original heading, rolls in a complete circle, all points of which are equidistant from an imaginary line extending forward from the starting point.

Base leg—A part of the airport traffic pattern. A flight path at right angles to the runway, following the downwind leg and followed by the final approach.

Basic operating weight—The weight of an aircraft including fuel and equipment necessary for flight.

Basket—The gondola of a balloon, regardless of the material of which it is made.

Beechcraft Baron (courtesy of Beech).

BC—Backcourse.

BC—Beginning climb.

BCKG—Backing.

BCM—Become.

BCN—Beacon.

BCST—Broadcast.

BCSTN—Broadcast station.

BD—Blowing dust (in sequence reports.)

BD—Beginning descent.

BDR—Border.

Beacon—A fixed reference point in aviation navigation. It may be visual, such as the rotating beacon at an airport, or it may be electronic such as non-directional homing beacons, outer markers, etc.

Bearing—The horizontal direction of an aircraft to any point, usually measured clockwise in 360 degrees relative to true or magnetic north. When navigating with VORs the term bearing is usually used with reference to the direction *to* the station, while the direction *from* the station is referred to as the radial.

Beaufort scale—A convenient scale for roughly indicating wind velocity in terms of a simple number (1 through 12) plus a descriptive word (calm through hurricane).

Becker Flugfunkwerk GmbH—Manufacturers of avionics equipment. (D757 Baden-Baden, West Germany).

Beech Aircraft Corporation—Manufacturers of a wide variety of general aviation aircraft from simple singles through turboprops. (9709 East Central, Wichita, KS 67201.)

Beechcraft—An aircraft manufactured by Beech Aircraft Corporation.

Bellanca Aircraft Corporation—Manufacturers of sport aircraft and high-performance single-engine aircraft. (P.O. Box 624, Alexandria, MN 56308.)

Beechcraft (courtesy of Beech).

Beech Aircraft Corporation single-engine piston aircraft specifications.

BEECH AIRCRAFT CORPORATION MODEL:	SUNDOWNER 180	SIERRA 200	BONANZA F33A	BONANZA V35B	BONANZA A36	BONANZA A36TC
ENGINE	L/IO-360-A4K	L/IO-360-A1B6	C/IO-520-BB	C/IO-520-BB	C/IO-520-BB	C/TSIO-520-UB
TBO (hours)	2,000	1,600	1,500	1,500	1,500	NA
PROPELLER	fixed pitch	const. speed	const. speed	const. speed	const. speed	const. speed
number of blades	2	2	2 or 3	2 or 3	2 or 3	3
LANDING GEAR	fixed	retractable	retractable	retractable	retractable	retractable
SEATS	4	4 to 6	4 to 5	4 to 5	4 to 6	4 to 6
TAKEOFF ground roll (ft)	1,130	1,185	1,002	1,002	1,140	847/1,176 (15%/0 flaps)
50' obstacle (ft)	1,955	1,660	1,769	1,769	2,040	1,758/2,012 (15%/0 flaps)
RATE OF CLIMB (fpm)	792	927	1,167	1,167	1,030	1,165
Vx (knots)	69	71	83	83	84	85
Vy (knots)	75	85	96	96	96	110
SERVICE CEILING (ft)	12,600	15,385	17,858	17,858	16,600	25,000
MAX SPEED (knots)	123 (sea level)	142 (sea level)	182 (sea level)	182 (sea level)	179 (sea level)	214 (10,000 ft)
CRUISE SPEED (knots)	84%:123 (4,500 ft)	75%:137 (10,000 ft)	75%:172 (6,000 ft)	75%:172 (6,000 ft)	75%:168 (6,000 ft)	75%:175 (10,000 ft)
economy (knots)	59%:98 (4,500 ft)	55%:115 (10,000 ft)	55%:157 (12,000 ft)	55%:157 (12,000 ft)	55%:150 (12,000 ft)	56%:170 (25,000 ft)
RANGE w. res. standard tanks (nm)	75%:533 (8,500 ft)	75%:646 (10,000 ft)	75%:378 (6,000 ft)	75%:378 (6,000 ft)	75%:365 (6,000 ft)	75%:322 (10,000 ft)
long-range tanks (nm)	NA	NA	75%:716 (6,000 ft)	75%:716 (6,000 ft)	75%:697 (6,000 ft)	75%:635 (10,000 ft)
economy, standard tanks (nm)	59%:597 (4,500 ft)	55%:686 (10,000 ft)	55%:439 (12,000 ft)	55%:439 (12,000 ft)	55%:405 (12,000 ft)	56%:314 (25,000 ft)
long range tanks (nm)	NA	NA	55%:838 (12,000 ft)	55%:838 (12,000 ft)	55%:790 (12,000 ft)	56%:730 (25,000 ft)
FUEL FLOW (pph)	75%:64.8	75%:61.2	75%:91.4	75%:91.4	75%:91.4	75%:100.8
economy (pph)	59%:48	55%:48	55%:69	55%:69	55%:69	56%:73.8
STALL clean/dirty (knots)	NA/51	65/60	64/51	64/51	62/52	67/57
LANDING ground roll (ft)	703	816	763	763	840	721
50' obstacle (ft)	1,484	1,462	1,324	1,324	1,450	1,449
RAMP WEIGHT (lbs)	2,455	2,758	3,412	3,412	3,612	3,666
TAKEOFF/LANDING WEIGHT (lbs)	2,450	2,750	3,400	3,400	3,600	3,650
USEFUL LOAD (lbs)	975	1,062	1,299	1,318	1,451	1,404
FUEL standard tanks (lbs)	342	342	264	264	264	264
long-range tanks (lbs)	NA	NA	444	444	444	444
WING LOADING (lbs per sq ft)	16.78	18.84	18.8	18.8	19.9	20.2
WING AREA (sq ft)	146	146	181	181	181	181
LENGTH/HEIGHT/SPAN external (ft)	25.75/8.25/32.75	25.75/8.08/32.75	26.71/8.45/33.5	26.45/7.55/33.5	27.5/8.45/33.5	27.5/8.45/33.5
LENGTH/HEIGHT/WIDTH cabin (ft)	7.9/4.3/3.67	7.9/4.3/3.67	10.02/4.22/3.5	10.02/4.22/3.5	10.95/4.22/3.5	10.95/4.22/3.5
PRICE standard tanks (1979 $s)	35,750	49,000	82,150	88,000	88,000	98,800
long-range tanks (1979 $s)	NA	NA	83,434	89,284	89,284	100,084
STANDARD EQUIPMENT:						
IFR PANEL	No	No	Yes	Yes	Yes	Yes
VHF NAV and COM w. OBI	No	No	Yes	Yes	Yes	Yes
						TURBOCHARGER

Beech Aircraft Corporation twin-engine piston aircraft specifications.

BEECH AIRCRAFT CORPORATION MODEL:	DUCHESS 76	BARON B55	BARON E55	BARON 58	BARON 58TC	BARON 58P	DUKE 60
ENGINE	L/O-360 (counter rot)	C/IO-470-L	IO-520-CB	C/IO-520-CB	C/TSIO-520-WB	C/TSIO-520-WB	L/TIO-541-E1C4
TBO (hours)	2,000	1,500	1,500	1,500	1,400	1,400	1,600
PROPELLERS	const. speed, ff	const. speed, ff	const. speed, ff	const. speed, ff	const. speed, ff	const. speed, ff	const. speed, ff
SEATS	4	4 to 6	4 to 6	6	4 or 6	4 or 6	4 or 6
number of blades, each	2	2 or 3	2 or 3	2 or 3	3	3	3
TAKEOFF ground roll (ft)	1,017	1,400	1,315	1,336	1,555	1,555	2,075
50 obstacle (ft)	2,119	2,154	2,050	2,101	2,643	2,643	2,626
RATE OF CLIMB (fpm)	1,248 (1 eng. 235)	1,693 (1 eng. 397)	1,682 (1 eng. 388)	1,660 (1 eng. 390)	1,418 (1 eng. 270)	1,481 (1 eng. 270)	1,601 (1 eng. 307)
Vx (knots) (IAS)	69	83	86	86	95	95	99
Vsse (knots) (IAS)	85	91	96	96	102	102	100
Vy (knots) (IAS)	85	107	101	104	115	115	120
Vyse (knots) (IAS)	85	99	99	100	115	115	110
SERVICE CEILING (ft)	19,650 (1 end. 6,170)	19,300 (1 end. 6,400)	19,100 (1 end. 6,600)	18,600 (1 end. 7,000)	25,000 (1 end. 13,490)	25,000 (1 end. 13,490)	30,000 (1 end. 15,100)
MAX SPEED (knots)	171	201	208	208	261 (25,000 ft)	261 (25,000 ft)	246 (18,000 ft)
CRUISE hi speed (knots)	75% 166 (6,000 ft)	75% 188 (6,000 ft)	77% 200 (6,000 ft)	77% 200 (6,000 ft)	77% 232 (20,000 ft)	77% 244 (25,000 ft)	74% 233 (25,000 ft)
economy (knots)	55% 151 (12,000 ft)	55% 173 (12,000 ft)	56% 184 (12,000 ft)	56% 184 (12,000 ft)	53% 194 (20,000 ft)	53% 202 (25,000 ft)	63% 217 (25,000 ft)
RANGE hi speed (nm)	75% 623 (6,000 ft)	75% 548 (6,000 ft)	77% 506 (6,000 ft)	77% 727 (6,000 ft)	77% 813 (20,000 ft)	77% 852 (25,000 ft)	74% 568 (25,000 ft)
economy (nm)	55% 780 (12,000 ft)	55% 688 (12,000 ft)	56% 610 (12,000 ft)	77% 877 (12,000 ft)	53% 958 (20,000 ft)	77% 1,008 (25,000 ft)	74% 1,072 (25,000 ft)
long range tanks hi speed (nm)	NA	75% 798 (6,000 ft)	77% 933 (6,000 ft)	77% 1,108 (6,000 ft)	53% 1,028 (20,000 ft)	53% 1,032 (25,000 ft)	63% 608 (25,000 ft)
long range tanks economy (nm)	NA	55% 991 (12,000 ft)	56% 1,135 (12,000 ft)	56% 1,339 (12,000 ft)	53% 1,217 (20,000 ft)	53% 1,229 (25,000 ft)	63% 1,168 (25,000 ft)
FUEL FLOW per engine hi speed (pph)	75%: 67.8	75%: 82	77%: 93	77%: 93	77%: 115.2	77%: 111.2	74%: 123.3
economy (pph)	55%: 46.8	56%: 60	56%: 69	56%: 93	53%: 73.8	53%: 74.4	63%: 105.1
STALL clean/dirty (knots) (IAS)	70/60	79/73	83/73	84/74	84/78	84/78	81/73
LANDING ground roll (ft)	1,000	1,467	1,237	1,439	1,378	1,378	1,318
50 obstacle (ft)	2,450	2,148	2,202	2,498	2,427	2,427	3,065
ACCELERATE-STOP DISTANCE (ft)	3,500	3,380	3,200	3,125	3,300	3,300	3,460
RAMP WEIGHT (lbs)	3,916	5,121	5,324	5,424	6,240	6,200	6,819
ZERO FUEL WEIGHT (lbs)	NA	NA	NA	NA	NA	NA	6,775
TAKEOFF/LANDING WEIGHT (lbs)	3,900	5,100	5,300	5,400	6,200	6,200	6,775
USEFUL LOAD (lbs)	1,470	1,895	2,059	2,071	2,452	2,241	2,436
FUEL standard tanks (lbs)	600	600	600	816	996	996	1,212 or 1,392
long range tanks (lbs)	NA	NA	816 or 996	996 or 1,164	1,140	NA	NA
WING LOADING (lbs per sq ft)	21.5	25.6	26.6	27.1	33	33	31.8
WING AREA (sq ft)	181	199.1	199.1	199.2	188.1	188.1	212.9
LENGTH/HEIGHT/SPAN exterior (ft)	29.9.5/8	28/9.53/37.83	28/9.9.5/37.83	29/9.5/37.78	28.78/9.22/37.78	28.78/9.22/37.78	33.78/12.33/39.25
LENGTH/HEIGHT/WIDTH cabin (ft)	7.9/4.3.67	10.0/8/4.17/3.5	11.7.54/17/3.5	12.6/4.22/3.5	12.6/4.22/3.5	12.6/4.22/3.5	11.8/4.3/4.17
TURBOCHARGER make	NA	NA	NA	NA	Garrett	Garrett	Garrett
PRESSURIZATION (psi)	NA	NA	NA	NA	NA	3.9	4.6
PRICE standard tanks (1979 $s)	91,850	128,850	157,850	183,750	206,650	250,000	313,000
long range tanks (1979 $s)	NA	130,480	159,480 or 160,880	185,800 or 190,500	211,350	254,700	317,000 or 321,775
STANDARD EQUIPMENT:							
IFR PANEL	No	Yes	Yes	Yes	Yes	Yes	Yes
VHF NAV and COM w. OBI	No	Yes	Yes	Yes	Yes	Yes	No
MARKER	No	No	No	No	No	No	No
GLIDE SLOPE	No	No	No	No	No	No	No
AUDIO PANEL	No	No	No	No	No	No	No
TRANSPONDER	No	No	No	No	No	No	Yes
ADF	No	No	No	Yes	No	yes	Yes

Beech Aircraft Corporation turboprop aircraft specifications.

BEECH AIRCRAFT CORPORATION MODEL:	KING AIR C90	KING AIR E90	KING AIR A100	KING AIR B100	SUPER KING AIR 200
ENGINES	PW/PT6A-21	PW/PT6A-28	PW/PT6A-28	GA/TPE 331-6-252B	PW/PT6A-41
shp (each)	550 (flat rated)	550 (flat rated)	680	715 (flat rated)	850 (flat rated)
TBO (hours)	3,500	NA	3,500	3,000	2,500
SEATS crew + passengers	2 + 4 to 8	2 + 4 to 8	2 + 6 to 13	2 + 6 to 13	2 + 6 to 13
TAKEOFF ground run (ft)	1,629	2,024	1,855	1,755	1,856
50 ft obstacle (ft)	2,261	1,553	2,681	2,694	2,579
RATE OF CLIMB (fpm)	1,955 (1 end. 539)	1,870 (1 end. 470)	1,963 (1 eng. 452)	2,139 (1 eng. 501)	2,450 (1 eng. 740)
Vsse (knots) (IAS)	100	99	113	111	115
Vyse (knots) (IAS)	107	110	119	125	121
Vmca (knots) (IAS)	90	86	85	85	86
ACCELERATE-STOP DISTANCE (ft)	3,498	3,736	4,275 (w. flaps 3,877)	3,923 (w. flaps 3,498)	3,364 (w. flaps 3,411)
SERVICE CEILING (ft)	28,100 (1 end. 15,050)	27,620 (1 eng. 14,390)	24,850 (1 eng. 9,300)	29,100 (1 end. 12,120)	31,000 (1 end. 19,150)
MAX SPEED (knots)	222 (12,000 ft)	249 (12,000 ft)	248 (10,000 ft)	268 (12,000 ft)	289
BEST CRUISE SPEED (knots)	222 (12,000 ft)	249 (12,000 ft)	248 (10,000 ft)	265 (12,000 ft)	285 (18,000 ft)
LONG RANGE CRUISE SPEED (knots)	191 (21,000 ft)	197 (21,000 ft)	207 (21,000 ft)	241 (21,000 ft)	232 (31,000 ft)
RANGE W. res. best speed (knots)	957 (12,000 ft)	1,004 (12,000 ft)	900 (10,000 ft)	1,015 (12,000 ft)	1,190 (18,000 ft)
long range speed (knots)	1,281 (21,000 ft)	1,625 (21,000 ft)	1,340 (21,000 ft)	1,325 (21,000 ft)	1,887 (31,000 ft)
FUEL FLOW per eng best speed (pph)	257	354	378	335	385
economy (pph)	162.9	172	204.5	235	195
STALL clean/dirty/100% flaps (knots IAS)	89/81/76	86/80/77	89/80/75	93/86/83	99/85/75
LANDING ground roll (ft)	1,075	1,030	1,390 w. prop rev.	1,290 w. prop rev.	1,120 w. prop rev.
50 ft obstacle (ft)	2,010	2,110	2,109 w. prop rev.	2,679 w. prop rev.	2,070 w. prop rev.
RAMP WEIGHT (lbs)	9,705	10,160	11,568	11,875	12,590
ZERO FUEL WEIGHT (lbs)	NA	NA	9,600	9,600	10,400
TAKEOFF/LANDING WEIGHT (lbs)	9,650/9,168	10,100/9,700	11,500/11,210	11,800/11,210	12,500/12,500
USEFUL LOAD (lbs)	3,933	4,108	4,771	4,774	3,647
FUEL CAPACITY (lbs)	2,572.8	3,175.8	3,149	2,149	3,644.8
WING LOADING (lbs per sq ft)	32.8	34.4	40.8	42.4	41.3
WING AREA (sq ft)	293.94	293.94	279.7	279.7	303
LENGTH/HEIGHT/SPAN external (ft)	35.5/14.24/50.25	35.5/14.24/50.25	39.95/15.35/45.85	39.95/15.35/45.85	43.75/15/54.5
LENGTH/HEIGHT/WIDTH cabin (ft)	12.67/4.75/4.5	12.67/4.75/4.5	16.67/4.75/4.5	16.67/4.75/4.5	16.67/4.75/4.5
PRESSURIZATION (psi)	4.6	4.6	4.6	4.6	6
PRICE (1979 $s)	630,500	820,750	907,500	935,000	1,174,000
STANDARD EQUIPMENT:					
VHF NAV and COM w. OBI, AUDIO PANEL	Yes	Yes	Yes	Yes	Yes
MARKER	Yes	Yes	Yes	Yes	Yes
GLIDE SLOPE	Yes	Yes	Yes	Yes	Yes
TRANSPONDER	Yes	Yes	Yes	Yes	Yes
ADF	Yes	No	Yes	Yes	Yes
DME	No	No	Yes	Yes	Yes
RMI	No	No	Yes	Yes	Yes
WEATHER RADAR	No	No	Yes	Yes	Yes
De-ICING wings/props	Yes	Yes	Yes	Yes	Yes
TOILET	No	Yes	Yes	Yes	Yes

Bell Helicopter Textron—Manufacturers of a variety of civilian and military helicopters. (P.O. Box 482, Fort Worth, TX 76101.)

Belly landing—An emergency or forced landing made without extending the landing gear. Belly landings are often made inadvertently. Usually damage, except to the propeller, is light. A proficient pilot, flying an aircraft with a two-bladed propeller and making an intentional gear-up landing, can stop the prop in a horizontal position, thus avoiding such damage.

Below minimums—Weather conditions below the minimums in terms of ceiling and/or visibility, as prescribed by the FARs for a particular type of activity such as landing, takeoff, VFR or IFR.

Bendix Avionics Division—Manufacturers of a wide variety of avionics instrumentation. (P.O. Box 9414, Ft. Lauderdale, FL 33310.)

Bendix Flight Systems Division—Manufacturers of long-range navigation systems. (Teterboro, NJ 07608.)

Bendix Instruments and Life Support Division—Manufacturers of encoding altimeters. (P.O. Box 4508, Davenport, IA 52808.)

Bernoulli's principle—The basic theory in describing lift: As velocity of a fluid (air) increases, the pressure decreases.

Best angle of climb—The combination of airspeed and power which enables an airplane to gain maximum altitude over the shortest distance. Also called *steepest angle of climb.*

Best rate of climb—The combination of airspeed and power which produces maximum gain in altitude within a given period of time. The best rate of climb produces a somewhat shallower climb angle than does the best angle of climb.

Bell 222 twin-turbine helicopter (courtesy of Bell).

Bell Helicopter Textron turbine helicopters specifications.

BELL HELICOPTER TEXTRON MODEL:	JETRANGER III	LONG RANGER	205A-1	222	212	214B-1
ENGINE	AL/250/C20B	AL/250-C20B	L/T53-13B	L/LTS101-650C (2)	PW/PT6T3 (2)	L/T550-8D
shp (each)	420	500	1,400	615	900	2,930
SEATS	5	7	11	7	15	16
RATE OF CLIMB max at sea level (fpm)	1,260	1,650	1,680	1,730	1,420	2,200
vertical (fpm)	300	320	850	NA	NA	NA
SERVICE CEILING (ft)	13,500	13,800	14,700	14,300	13,200	20,000
HIGE (ft)	12,400	13,200	10,400	10,300	11,000	15,000
HOGE (ft)	8,800	8,100	6,000	6,400	9,300	13,400
MAX SPEED sea level (knots)	130	130	115	144	104	146
NORMAL CRUISE SPEED (knots)	118 (sea level)	113 (sea level)	110 (sea level)	143 (sea level)	100 (sea level)	140 (sea level)
FUEL CAPACITY (lbs)	509	657	2,674	1,266	1,440	1,367
RANGE (nm)	300	317	275	386	230	196
RANGE w. max payload (nm)	230	123	137	305	230	196
GROSS WEIGHT (lbs)	3,200	4,050	9,500	7,650	11,200	13,800
EMPTY WEIGHT (lbs)	1,580	2,156	5,988	5,247	6,143	7,813
USEFUL LOAD (lbs)	1,620	1,894	3,512	2,403	5,057 (sling 5,000)	5,987 (sling 8,000)
PAYLOAD w. full fuel (lbs)	911	1,037	638	737	3,217	4,420
LENGTH/HEIGHT/WIDTH external (ft)	39.1/9.5/6.4	42.75/11.75/7.4	57.1/14.4/9.1	48/11.2/9	57.17/12/67/NA	60.75/13.83/NA
cabin (ft)	6.9/4.3/4	8.3/4.2/4.2	11.3/4.1/8	7.5/4.3/4.8	11.3/4.1/NA	NA
MAIN ROTOR number of blades	2	2	2	2	2	2
diameter (ft)	33.33	37	48	39	48	50
TAIL ROTOR number of blades	2	2	2	2	2	2
diameter (ft)	5.5	5.4	8	6.5	NA	NA
PRICE (1979 $s)	225,000	362,000	825,000	975,000	1,040,000	1,425,000

Bellanca Aircraft Corporation single-engine piston aircraft specifications.

BELLANCA AIRCRAFT CORPORATION MODEL:	CITABRIA 115 7ECA	DECATHLON FP 8KCAB	SUPER VIKING 17-30A	TURBO VIKING 17-31ATC	ARIES T-250
ENGINE	L/O-235-K2C	L/AEIO-320-E2B	C/IO-520-K	L/TIO-540-S1AD	L/O-540-A4D5
TBO (hours)	2,000	2,000	1,500	2,000	2,000
PROPELLER	fixed pitch	fixed pitch	const. speed	const. speed	const. speed
number of blades	2	2	2	3	2
LANDING GEAR	fixed, tail wheel	fixed, tail wheel	retract. tricycle	retract. tricycle	retract. tricycle
SEATS	2	2	4	4	4
TAKEOFF ground roll (ft)	340	630	510	510	NA
50' obstacle (ft)	716	1,180	750	750	NA
RATE OF CLIMB (fpm)	725	1,000	1,210	1,170	1,240
Vx (knots)	50.4	55.7	65.2	65	NA
Vy (knots)	65.2	66.1	95.7	95.7	NA
SERVICE CEILING (ft)	12,000	16,000	20,000	24,000	18,100
MAX SPEED (knots)	108.7 (sea level)	127.8 (sea level)	180.9 (sea level)	193 (24,000 ft)	187 (sea level)
CRUISE SPEED (knots)	75%:107	75%:119 (7,500 ft)	75%:175.7 (7,500 ft)	75%:193 (7,500 ft)	75%:181 (7,500 ft)
economy (knots)	55%:96	55%:105 (7,500 ft)	55%:169.6 (7,500 ft)	55%:144 (7,500 ft)	55%:174 (7,500 ft)
RANGE w. res. (nm)	75%:319 (7,500 ft)	75%:468 (7,500 ft)	75%:826 (7,500 ft)	75%:666 (7,500 ft)	75%:990 (7,500 ft)
economy (nm)	55%:528 (7,500 ft)	55%:540 (7,500 ft)	55%:930 (7,500 ft)	55%:695 (7,500 ft)	55%:1,170 (7,500 ft)
FUEL FLOW (pph)	75%:48	75%:52	75%:96.2	75%:94.8	75%:84
economy (pph)	55%:30	55%:41	55%:70.5	55%:69.8	55%:72
STALL clean/dirty (knots)	36/NA	37/NA	63/56	63/56	63/55.6
LANDING 50' obstacle (ft)	890	1,450	1,420	1,440	NA
RAMP WEIGHT (lbs)	1,650	1,800	3,325	3,325	3,150
TAKEOFF/LANDING WEIGHT (lbs)	1,650	1,800	3,325	3,325	3,150
ZERO FUEL WEIGHT (lbs)	1,067	1,260	2,185	2,372	1,850
USEFUL LOAD (lbs)	583	540	1,140	1,053	1,300
FUEL CAPACITY (lbs)	210	240	408	408	456
WING LOADING (lbs per sq ft)	10	10.6	20.6	20.6	18.5
WING AREA (sq ft)	NA	NA	NA	NA	170
LENGTH/HEIGHT/SPAN external (ft)	22.7/7.7/33.4	22.9/7.7/32	26.3/7.3/34.2	26.3/7.3/34.2	26.2/8.6/31.4
PRICE (1979 $s)	17,900	23,900	56,900	70,380 DUAL RAJAY TURBOS	NA

BFA—Balloon Federation of America.

BFDK—Before dark.

BFR—Before.

BFO—Beat frequency oscillator.

BGN—Begin; began.

BHND—Behind.

BHP—Brake horse power.

BINOVC—Breaks in the overcast.

Bird—Slang for airplane.

Birdman—Prior to World War I, the common term for flyers.

Bit—A piece of information stored in a computer memory.

BL—Between layers.

Black box—Aviation jargon for any piece of avionics equipment.

Blade—Part of a propeller. Propellers may have two or more blades. Each blade is, in fact, an airfoil, twisted lengthwise to compensate for the increasing distance from the hub.

Blade angle—The angle of the propeller blade relative to the plane of rotation. It is generally measured at a point three quarters of the distance from the hub. On fixed-pitch propellers the angle is constant. Variable-pitch propellers can be increased or decreased by the pilot during flight.

Blade coning—The upward bending of helicopter rotor blades when they are in the process of producing lift.

Blast fence—A barrier erected on airports near the takeoff positions to divert or dissipate jet or propeller blast.

Blast valve—The valve on the burner of a hot-air balloon with which the pilot can regulate the amount of heat directed into the balloon.

BLDG—Building.

Blimp—A non-rigid airship using gas for lift.

Blind flying—Flying by instruments alone.

Blind speed—The rate of closing or departure of a target relative to the radar antenna at which cancellation of the primary radar target by moving target indicator circuits (MTI) causes reduction or complete loss of the signal.

Blind spot—Areas from which radio or radar transmissions cannot be received. Also portions of an airport invisible from the tower.

Blind zone—See blind spot.

Blip—The reflected or transponder-augmented echo from an aircraft or other object and seen on the radar scope as a spot of light. It indicates the position of the object relative to the location of the radar scope.

BLN—Balloon.

BLO—Below.

Blower—A portable fan equipped with a small gasoline motor, used by balloonists to force air into the balloon as the first step toward inflation of a hot-air balloon.

Blower—Mechanically-driven supercharger.

Blimp.

BLZD—Blizzard.

BMEP—Brake mean effective pressure.

BMEWS—Ballistic missile early warning system.

BN—Blowing sand (in sequence reports.)

BNDRY—Boundary.

BNTH—Beneath.

Boeing Vertol Company—Manufacturers of heavy-duty helicopters, the Chinook. (Boeing Center, P.O. Box 16858, Philadelphia, PA 19142).

Bonanza—A family of high-performance single-engine aircraft manufactured by Beech Aircraft Corporation.

Bonzer, Inc.—Manufacturers of radio altimeters. (90th and Cody, Overland Park, KS 66214.)

Boost pump—An electric fuel pump used to force increased fuel pressure during engine start.

Boundary lights—See runway lights.

BOVC—Base of overcast.

BOW—Basic operating weight.

BPT—Beginning procedure turn.

BRAF—Braking action fair.

BRAG—Braking action good.

Brain bag—Aviation jargon for the case in which the pilot carries his charts, computers, plotters and other necessary paraphernalia.

Brake horsepower—(BHP)—The usable horsepower delivered to the propeller shaft. The amount of energy remaining after the actual horse-

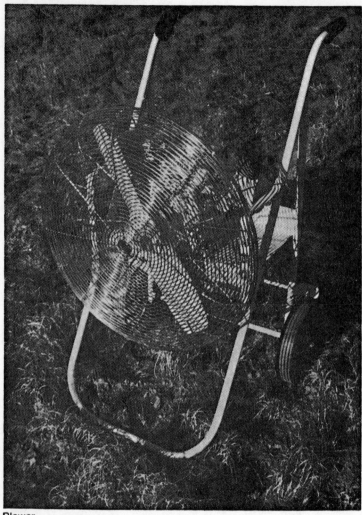

Blower.

power developed by the engine is reduced by the friction of moving parts and the amount dissipated by other engine-driven systems.

Brake mean effective pressure—The average of effective combustion pressures acting upon the crankshaft.

Braking action—The term used to describe the condition of a runway, usually in winter weather when snow and/or ice are present, affecting the ability of a landing aircraft to use its brakes in order to come to a full stop.

BRAN—Braking action nil.

Beechcraft Bonanza (courtesy of Beech).

Boeing Vertol Company heavy-duty twin-turbine helicopters specifications.

BOEING VERTOL COMPANY MODEL:	CHINOOK (Passenger)	CHINOOK (Utility)
ENGINES	L/AL 5512 (2)	L/AL 5512 (2)
shp (each)	4,075	4,075
SEATS crew plus passengers	2 + 44	2
(Performance figures at 47,000 lbs except where otherwise noted)		
RATE OF CLIMB sea level (fpm)	1,350 (cont. pwoer)	1,350 (cont. power)
vertical (fpm)	654 (t.o. power)	1,050 (t.o. power)
one engine out (fpm)	375 (30 min power)	375 (30 min power)
SERVICE CEILING (ft)	11,000 (1 eng. 2,650)	11,000 (1 eng. 2,650)
HIGE (ft)	9,150	10,350
HOGE (ft)	4,750	7,150
MAX SPEED sea level (knots)	145 (max cont. power)	145 (max cont. power)
NORMAL CRUISE SPEED (knots)	135	135
Vne (knots)	165	165
FUEL CAPACITY (lbs)	14,024	4,026
RANGE w full fuel (nm)	740	180
w max payload (nm)	660 (44 passengers)	40 (20,000 lbs ext. ld)
GROSS WEIGHT (lbs) internal load	47,000	47,000
external load	49,000	51,000
EMPTY WEIGHT (lbs)	24,449	20,323
USEFUL LOAD (lbs) internal load	22,551	26,677
external load	24,551	30,677
PAYLOAD w. full fuel (lbs) internal ld.	7,476	21,626
external ld.	10,070	26,194
PAYLOAD w. fuel for 100 nm, internal	17,755	22,364
external	19,133	26,310
FUEL w. max payload (lbs)	12,700 (44 pass.)	2,220 (28,000 ext. ld.)
LENGTH/HEIGHT/WIDTH external (ft)	90/18.6(60	90/18.6/60
cabin (ft)	30.17/6.5/8.25	NA
FWD. ROTOR number of blades	3	3
diameter (ft)	60	60
AFT ROTOR number of blades	3	3
diameter (ft)	60	60
IFR CERTIFICATION STATUS	pending	NA
PRICE (1979 $s)	NA	NA
STANDARD EQUIPMENT: VHF COM; RADAR ALTIMETER; AUTOPILOT.		

Brantly-Hynes piston helicopters specifications.

BRANTLY-HYNES HELICOPTER, INC. MODEL	B-2B	305
ENGINE	L/IOV-360-A1A	L/IOV-540-B1A
hp	180	305
TBO (hors)	1,000	800
SEATS	2	5
RATE OF CLIMB, max sea level (fpm)	1,330	975
vertical (fpm)	500	400
SERVICE CEILING (ft)	10,900	12,000
HIGE (ft)	6,700	4,000
HOGE (ft)	4,400	2,400
MAX SPEED sea level (knots)	87	104
NORMAL CRUISE SPEED (knots)	83	96
Vne (knots)	87	104
FUEL CAPACITY (lbs)	186	258
RANGE 3˙ full fuel (nm)	196	239
w. max payload	196	239
GROSS WEIGHT (lbs)	1,670	2,900
EMPTY WEIGHT (lbs)	1,000	1,700
PAYLOAD w. full fuel (lbs)	484	942
w. fuel for 100 nm (lbs)	574	1,080
FUEL w. max payload	186	258
LENGTH/HEIGHT/WIDTH external (ft)	28/6.8/5.7	32.9/8/6.9
cabin (ft)	5/3.2/3.9	6.7/4.7/4.5
MAIN ROTOR number of blades	3	3
diameter (ft)	23.7	38.5
TAIL ROTOR number of blades	2	2
diameter (ft)	4.2	4.2
IFR CERTIFICATION STATUS	No	No
PRICE (1979 $s)	49,950	89,950

Brantly-Hynes Helicopter, Inc.—Manufacturers of single-engine piston helicopters. (P.O. Box 1046, Frederick, OK 73542.)

BRAP—Braking action poor.

Brave—An agricultural aircraft manufactured by Piper Aircraft Corporation.

Bravo—In aviation phraseology the term used for the letter B.

Breaks in the overcast—Cloud conditions in which the cloud cover obscures 90 or more percent of the sky, but less than 100 percent.

Brelonix, Inc.—Manufacturers of HF transceivers and other avionics equipment. (106 North 36th Street, Seattle, WA 98103.)

BRF—Brief.

BRG—Bearing.

Briefing—Information given the pilot with reference to weather, NOTAMs and anything else required for the planning of a proposed flight.

Brittain Industries, Inc.—Manufacturers of autopilots. (P.O. Box 51370, Tulsa, OK 74151.)

BRK—Break.

Brantly-Hynes helicopter (courtesy of Brantly-Hynes).

BRKN—Broken.
Broadcast—A transmission of information requiring no acknowledgement by the pilot.
Broken cloud—See broken overcast.

Piper Brave (courtesy of Piper).

Broken overcast—Cloud cover which obscures the sky by between 60 and 90 percent.

BS—Standard broadcast station.

BS—Blowing snow (in sequence reports.)

BSFC—Basic specific fuel consumption.

BTN—Between.

BTR—Better.

BTU—British thermal unit. A unit used to measure heat output.

Bucket—The gondola of a balloon.

Buck the weather—A term used to describe flying into or through rough weather or to proceed despite adverse weather conditions.

Buffet—The shudder of an airframe caused by disturbed airflow set up by some part of the aircraft.

Buffeting—See *buffet.*

Bulkhead—A piece of the structure of an aircraft; a more or less circular section of aluminum or other rigid materal to which the side panels are attached.

Burble point—The angle of attack which results in air separation above the wing. Also called the critical or stalling angle of attack. At this angle of attack the wing loses its lift capability.

Burbling—The separation of the airflow from an airfoil, especially its upper surface, causing loss of lift and increased drag.

Bulkhead.

Burner.

Burner—The heater carried aloft in a hot-air balloon.

Bus bar—A section of the instrument panel containing switches and circuit breakers or fuses.

Bush pilot—A pilot operating in thinly or unpopulated areas, such as Alaska or Central or South America.

Business aviation—The use of aircraft in the pursuit of business. Also the title of a weekly newsletter published in Washington by Ziff-Davis Publishing Co.

Buys Ballot's law—Refers to the fact that if a person in the Northern Hemisphere stands with his back to the wind, his left hand will point to the area of low pressure.

BY—Blowing spray (in sequence reports.)

BYD—Beyond.

C

C—Calm (on sequence reports.)

C.—Celsius.

C.—Centigrade.

C—Charlie in the phonetic alphabet.

C—Circling approach (on approach charts.)

C—Control tower.

C—Central standard time.

C—Continental air mass.

CAA—Civil Aviation Administration, the forerunner to the FAA.

CAB—Civil Aeronautics Board; the government agency supervising airline activity.

CADIZ—Canadian Air Defense Identification Zone.

CAF—Cleared as filed.

Caging mechanism—A device which locks a gyro or compass into a desired position, controlled by a knob on the instrument face. Gyros should be uncaged only in the straight and level flight or on the ground.

Calculator—A mechanical or electronic instrument capable of making mathematical calculations. Small electronic pocket calculators should not be used in the cockpit while navigating with the ADF, as they tend to confuse the ADF readout.

Calibrated airspeed—(CAS)—Indicated airspeed corrected for instrument and installation error.

Call sign—The name, numbers and letters identifying an aircraft or a ground station, such as "Bonanza Three Two Six Eight Hotel", or "Santa Fe Radio".

Call-up—The initial voice contact between an aircraft and a facility, using the call sign of the aircraft or station being called and that of the caller, such as: "Santa Fe Radio, Bonanza Three Two Six Eight Hotel, over."

Calm—The absence of wind with speeds of more than 3 knots.

Camber—The curvature of a wing or other airfoil, measured from the leading to the trailing edge.

Cameron Balloons, Ltd.—A manufacturer of hot-air balloons located in England. (U.S. address: 3600 Elizabeth Road, Ann Arbor, MI 48103.)

Canadair, Ltd.—A Canadian airframe manufacturer, producer of the Challenger 600. (P.O. Box 6087, Montreal, Canada H3C 3G9.) See *Challenger*.

Canadian Marconi—Manufacturers of long-range navigation equipment. (2442 Trenton Avenue, Montreal, Canada H3P 1Y9.)

CAP—Civil Air Patrol.

Caproni Vizzola—An Italian manufacturer of high-performance sailplanes. (Milano, Italy.)

Carburetion—The act of mixing fuel and air to the proportion necessary for combustion in a piston-engine aircraft.

Carburetor—The device which measures the flow of air and fuel to a piston engine.

Carburetor heat system—A small heating unit located near the carburetor throat and controlled by a plunger, lever or other device on the instrument panel. It is used to melt ice accumulations in the carburetor. It should be turned full on at the first sign of carburetor-ice buildup causing a certain degree of power reduction. When the carburetor-heat system is turned on, it causes an enriching effect on the mixture and the mixture should therefore be appropriately leaned.

Carburetor ice—Ice forming in the carburetor.

Carburetor icing—The formation of ice in the throat of the carburetor when moist air, expanding while passing through it, cools quickly due to the vaporization of fuel and the moisture condenses as frost or ice. It initially reduces the power output and, if not stopped, may cause the engine to quit.

Cardinal—A family of single-engine fixed-gear and retractable-gear aircraft produced by Cessna Aircraft Company. Production of Cardinals was stopped in 1978.

Cessna Cardinal (courtesy of Cessna).

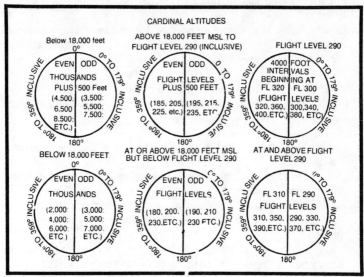

Cardinal altitudes (courtesy of FAA).

Cardinal altitudes—Specific altitudes or flight levels to be used by aircraft when operating under certain conditions or in given directions. Cardinal altitudes for VFR aircraft flying in an easterly direction (0 to 179°) are odd thousands plus 500 feet, starting at 3,000 feet agl. For VFR aircraft flying in a westerly direction (180 to 359°) they are even thousands plus 500 feet, starting at 3,000 feet agl. For IFR aircraft in uncontrolled airspace the cardinal altitudes are odd or even thousands depending on easterly or westerly direction of flight, also starting at 3,000 feet agl.

Cardinal heading—Any of the four major compass headings; north, south, east, west.

CAS—Calibrated airspeed.

CAT—Category landing.

CAT—Clear-air turbulence.

Category I landing—The standard ILS landing requiring a ceiling of 200 feet and a half mile visibility at airports equipped with a full ILS.

Category II landing—A low-minimum approach system using modified ILS ground equipment and appropriate cockpit instrumentation, permitting landings when the ceiling is down to 100 feet and visibility is at least 1,200 feet along the runway (Runway Visual Range). Pilots must be especially qualified to fly CAT II approaches.

Category III landing—A landing in zero-zero conditions, not authorized in the U.S. for civil aircraft except in test situations.

Caution area—Airspace within certain geographic limits in which military activities are conducted that are not hazardous but are of interest to non-participating pilots. Caution areas are designated on charts by the letter C.

Caution range—The range of airspeeds within which an aircraft should not be operated under conditions of other than smooth air. It is indicated on the face of the airspeed indicator by a yellow arc.

CAVU—Ceiling and visibility unlimited. Technically clear or scattered clouds and visibilities of better than 10 miles.

CB—Citizens band. A range of frequencies reserved for use by the average (appropriately licensed) citizen.

CDI—Course deviation indicator.

Ceiling—The altitude of the lowest layer of broken or overcast clouds not classified as thin or partial obscuration.

Ceiling—The highest altitude a specific aircraft can reach under standard atmospheric conditions.

Celestial navigation—The determination of the geographic position of an aircraft, using stars as reference points. Today, it is rarely used in aviation.

Celsius—A temperature scale in which zero represents the melting point of ice (the freezing point of water) and 100° the boiling point of water, both at sea level. It is equal to 5/9 minus 32° of the Fahrenheit scale. Also called Centigrade.

Center—Air route traffic control center.

Center area—The geographical limits of the air space controlled by a given ARTCC.

Centerline—The line, either actually visible or imaginary, running along the longitudinal center of a runway. The term is used also in relation to radio ranges, airways, and aircraft fuselages.

Centerline thrust—The thrust produced by both engines of a twin-engine aircraft with a puller engine in front and a pusher engine in back, such as the Cessna Skymaster.

Center of gravity—(CG)—The point at which the moments in all direction are zero. The geometric center of balance.

Centigrade—See Celsius.

Centimeter—1/100 meter (See conversion charts.)

Centralized control system—A computerized electronic system permitting the pilot to control all his avionics equipment through one central control panel.

Centrifugal force—The reaction of a body against a force causing it to move in a curved path, caused by inertia.

Centurion—A family of high-performance single-engine aircraft produced by Cessna Aircraft Company. The Pressurized Centurion was the first successful application of pressurization in a single-engine aircraft. (Previously Mooney Aircraft Corporation built a relatively unsuccessful pressurized single.)

Century I, II, III, IV—A family of autopilots manufactured by Edo-Aire/ Mitchell.

Cessna Centurion (courtesy of Cessna).

Certificate—A document, generally issued by the FAA, certifying that a pilot, mechanic, aircraft, new design or other person or piece of equipment meets the standards set down for him, her or it.

Certification—The process of obtaining a certificate. Most frequently used with reference to new types of aircraft.

Cessna Aircraft Company—Manufacturers of a full line of aircraft, from two-seat trainers to executive jets. (Box 1521, Wichita, KS 67201.)

CFIT—Controlled flight into terrain.

CFN—Confine.

CG—Center of gravity.

CG range—The distance along the longitudinal axis of an airplane within which the CG must fall in order for the aircraft to be properly balanced. Aircraft loaded beyond the CG range may become uncontrollable.

Chaff—Narrow strips of metallic reflectors, used to reflect radar energy. When dropped from an aircraft they tend to produce a large target on the radar screen.

Challenger 600—A wide-body executive jet aircraft, the first of its kind, produced by Canadair, Ltd.

Champion—A family of sport aircraft produced by Bellanca Aircraft Corporation.

Champion Decathlon—One of the Champion family of sport aircraft.

Chancellor—A pressurized twin-engine piston aircraft manufactured by Cessna Aircraft Company.

Chandelle—An aerobatic maneuver to gain maximum altitude in a minimum distance and time, while reversing the direction of flight by 180 degrees. It starts with a dive to gain speed followed by a well-coordinated climbing turn ending in level flight, at minimum controllable airspeed, in the opposite direction.

Changeover point—The point of which a pilot switches from one VOR, VORTAC or DME station to the next, usually located approximately halfway between stations and assures reliable reception distance.

Chaparral—A high-performance single-engine aircraft produced by Mooney Aircraft Corporation.

Cessna Aircraft Company single-engine piston aircraft specifications (continued on page 75).

CESSNA AIRCRAFT COMPANY MODEL	152/II/AEROBAT	SKYHAWK/II	HAWK XP/II	180 SKYWAGON/II	185 SKYWAGON/II	SKYLANE/II	SKYLANE RG/II
ENGINE	L/O-235-L2C	L/O-320-H2AD	C/IO-360-K	C/IO-470-U	C/IO-520-D	C/O-470-U	L/O-540-J3C5D
TBO (hours)	2,000	2,000	1,500	1,500	1,500	1,500	2,000
PROPELLER	fixed pitch	fixed pitch	const. speed	const. speed	const. speed	const. speed	const. speed
number of blades	2	2	2	2	2	2	2
LANDING GEAR	fixed	fixed	fixed	fixed	fixed	fixed	retractable
SEATS	2	4	4	4 to 6	6	4	4
TAKEOFF ground roll (ft)	725	805	800	625	770	705	820
50' obstacle (ft)	1,340	1,440	1,270	1,205	1,365	1,350	1,570
RATE OF CLIMB (fpm)	715	770	870	1,100	1,010	1,010	1,140
Vx (knots)	55	59	59	61	73	54	64
Vy (knots)	67	73	81	81	88	78	88
SERVICE CEILING (ft)	14,700	14,200	17,000	17,700	17,150	16,500	14,300 (18,000 w. EGT)
MAX SPEED (knots)	110 (*109)	125	133	148	155	148	160
CRUISE SPEED (knots)	75%:107 (*106)	75%:122	80%:130	75%:142	75%:145	75%:144	75%:156
economy (knots)	55%:91	55%:115	60%:116	55%:124	55%:129	55%:127	55%:139
RANGE w. res. standard tanks (nm)	75%:350 (*345)	75%:485	80%:480	75%:825	75%:825	75%:825	75%:890
long-range tanks (nm)	75%:580 (*575)	75%:630	80%:675	NA	NA	NA	NA
economy, stand. tanks (nm)	55%:415 (*410)	55%:575	55%:575	55%:1,010	55%:835	55%:1,095	55%:1,135
long-range tanks (nm)	55%:690 (*680)	55%:750	55%:815	NA	NA	NA	NA
FUEL FLOW (pph)	75%:43.2	75%:58.5	75%:79.5	75%:85.4	75%:107.2	75%:85.2	75%:91
economy (pph)	55%:28.3	55%:42.1	55%:48.2	55%:54.8	55%:68.1	55%:54.6	55%:58.7
STALL clean/dirty (knots)	48/43	50/44	53/46	53/48	56/49	54/50	54/50
LANDING ground roll (ft)	475	520	620	480	480	590	600
50' obstacle (ft)	1,200	1,250	1,270	1,365	1,400	1,350	1,320
RAMP WEIGHT (lbs)	1,675	2,307	2,558	2,810	3,362	2,960	3,112
TAKEOFF/LANDING WEIGHT (lbs)	1,670	2,300	2,550	2,800	3,350	2,950	3,100
EMPTY WEIGHT (lbs)	1,101/1,146 (*1,132)	1,397/1,444	1,541/1,588	1,663/1,684	1,669/1,719	1,690/1,760	1,720/1,771
USEFUL LOAD (lbs)	574/529 (*543)	910/863	1,017/970	1,167/1,116	1,681/1,631	1,260/1,190	1,380/1,329
FUEL standard tanks (lbs)	147	240	294	504	504	528	528
long-range tanks (lbs)	225	300	396	NA	NA	NA	NA
MAX PAYLOAD (lbs)	460	800	880	1,190	1,190	880	880
PAYLOAD w. full fuel, stand. tanks (lbs)	427/395 (*396)	670/643	723/695	664/612	1,177/1,127	772/718	852/801
long-range tanks (lbs)	349/317 (*318)	610/583	621/593	NA	NA	NA	NA
FUEL w. max payload (lbs)	114/82 (*83)	110/63	137/109	not possible	491/441	370/300	500/449
WING LOADING (lbs per sq ft)	10.5	13.2	14.7	16.1	19.3	16.9	17.8
WING AREA (sq ft)	159.5	174	174	174	174	174	174
LENGTH/HEIGHT/SPAN external (ft)	24.04/8.5/33.22	26.96/8.76/35.85	27.22/8.76/35.85	25.6/7.75/35.85	25.6/7.75/35.85	28.9/9.25/35.85	28.6/8.95/35.85
LENGTH/HEIGHT/WIDTH cabin (ft)	7.92/3.75/3.08	9/4/3.33	9/4/3.33	11.67/3.92/3.35	11.67/3.92/3.35	11.33/4.04/3.64	11.17/4.04/3.64
TURBOCHARGER make/type	NA	NA	NA	NA	NA	NA	NA
PRESSURIZATION (psi)	NA	NA	NA	NA	NA	NA	NA
PRICE standard tanks (1979 $s)	16,950 to 23,100	25,950 to 33,150	33,950 to 41,150	37,650 to 47,850	45,350 to 55,550	39,995 to 49,645	52,895 to 62,545
long-range taks (1979 $s)	add 435	add 435	add 435	NA	NA	NA	NA
STANDARD EQUIPMENT:							
IFR PANEL	II	Yes	Yes	Yes	Yes	Yes	Yes
VHF nav and com	II/*	II	II	II	II	II	II
AUDIO PANEL	No	No	No	II w. nav-pac	II w. nav-pac	II w. nav-pac	II w. nav-pac
MARKER	II w. nav-pac	II w. nav-pac	II w. nav-pac	II w. nav-pac	II w. nav-pac	II w. nav-pac	II w. nav-pac
GLIDE SLOPE	No	No	No	II w. nav-pac	II w. nav-pac	II w. nav-pac	II w. nav-pac
TRANSPONDER	No	II w. nav-pac	II w. nav-pac	II w. nav-pac	II w. nav-pac	II w. nav-pac	II w. nav-pac
ENCODING ALTIMETER	No	No	No	No	No	No	No
ADF	No	No	No	No	No	No	No
DME	No	No	No	No	No	No	No
RMI	No	No	No	No	No	No	No
RNAV	No	No	No	No	No	No	No
AUTOPILOT	No	No	No	No	No	No	No

Cessna Aircraft Company single-engine piston aircraft specifications (continued from page 74).

CESSNA AIRCRAFT COMPANY — MODEL:	STATIONAIR 6/II	TURBO STATIONAIR 6/11	STATIONAIR 7/II	TURBO STATIONAIR 7/II	CENTURION/II	TURBO CENTURION/II	PRESSURIZED CENTURION/II
ENGINE	C/IO-520-F	C/TSIO-520-M	C/IO-520-F	C/TIO-520-M	C/IO-520-L	C/TSIO-520-R	C/TSIO-520-P
TBO (hours)	1,500	1,400	1,500	1,400	1,500	1,400	1,400
PROPELLER	const. speed	const. speed	const. speed	const. speed	const. speed	const. speed	const. speed
number of blades	3	3	3	3	3	3	3
LANDING GEAR	fixed	fixed	fixed	fixed	retractable	retractable	retractable
SEATS	6	7	7	7	6	6	6
TAKEOFF ground roll (ft)	900	835	1,100	1,030	1,250	1,300	1,300
50 obstacle (ft)	1,780	1,640	1,970	1,860	2,030	2,160	2,160
RATE OF CLIMB (fpm)	920	1,010	810	885	950	930	930
Vx (knots)	66	55	70	74	79	82	80
Vy (knots)	84	88	87	87	96	100	100
SERVICE CEILING (ft)	14,600	27,000 (at 17,000 ft)	13,300	26,000 (at 17,000 ft)	17,300	27,000	23,000 (max operating)
MAX SPEED (knots)	156	174 (at 17,000 ft)	150	170 (at 17,000 ft)	175	204 (at 17,000 ft)	206 (at 17,000 ft)
CRUISE SPEED (knots)	75%/147	80%/167 (at FL 200)	75%/143	80%/161 (at FL 200)	75%/171	80%/196 (at FL 200)	80%/200 (at FL 200)
economy (knots)	55%/130 (at 10,000 ft)	55%/154 (at FL 200)	55%/126	55%/154 (at FL 200)	55%/154	55%/171 (at FL 200)	55%/175 (at FL 200)
RANGE w. res. standard tanks (nm)	75%/565	80%/350 (at FL 200)	75%/470	80%/525 (at FL 200)	75%/470	80%/815 (at FL 200)	80%/770 (at FL 200)
economy, stand. tanks (nm)	55%/690	55%/855 (at FL 200)	55%/390	55%/385 (at FL 200)	55%/610	55%/855 (at FL 200)	55%/925 (at FL 200)
long-range tanks (nm)	NA	NA	75%/690	80%/470 (at FL 200)	75%/855	80%/940 (at FL 200)	NA
economy, stand. tanks (nm)	NA	NA	55%/385	55%/385 (at FL 200)	55%/1,065	55%/940 (at FL 200)	NA
FUEL FLOW (pph)	75%/105.6	80%/120	75%/115.7	80%/135	75%/104.7	80%/121.4	80%/130.2
economy (pph)	55%/67.7	55%/72.3	55%/77.1	55%/82.6	55%/67.6	55%/72.2	55%/75.2
STALL clean/dirty (knots)	62/54	62/54	65/58	65/58	65/56	67/58	67/52
LANDING ground roll (ft)	735	735	765	765	765	765	765
50 obstacle (ft)	1,395	1,500	1,500	1,500	1,500	1,500	1,500
RAMP WEIGHT (lbs)	3,612	3,616	3,812	3,816	3,812	4,016	4,016
TAKEOFF/LANDING WEIGHT (lbs)	3,600	3,600	3,800	3,800	3,800	4,000	4,000
EMPTY WEIGHT (lbs)	1,907/1,968	1,968/2,028	2,064/2,133	2,141/2,210	2,117/2,182	2,205/2,271	2,318/2,383
USEFUL LOAD (lbs)	1,693/1,632	1,632/1,572	1,736/1,667	1,659/1,590	1,683/1,618	1,795/1,729	1,682/1,617
FUEL standard tanks (lbs)	528	528	324	324	534	534	534
long-range tanks (lbs)	NA	NA	438	438	NA	NA	NA
MAX PAYLOAD (lbs)	1,200	1,370	1,490	1,490	920	920	1,220
PAYLOAD w. full fuel, stand. tanks (lbs)	1,165/1,104	1,104/1,044	1,412/1,343	1,335/1,226	920/920	920/920	1,148/1,083
long-range tanks (lbs)	NA	NA	1,298/1,229	1,221/1,152	NA	NA	NA
FUEL w. max payload (lbs)	493/432	262/202	246/177	169/100	534/534	534/534	462/397
WING AREA (sq ft)	174	174	174	174	175	175	175
WING LOADING (lbs per sq ft)	20.7	20.7	21.8	21.8	21.7	22.9	22.9
LENGTH/HEIGHT/SPAN external (ft)	28.25/9.26/35.85	28.25/9.26/35.85	32.17/9.57/35.85	32.17/9.57/35.85	28.22/9.7/36.75	28.22/9.7/36.75	28.22/9.7/36.75
LENGTH/HEIGHT/WIDTH cabin (ft)	12.08/4.04/3.67	12.08/4.04/3.67	14/4/3.71	14/4/3.71	12.67/4.06/3.69	12.67/4.06/3.69	10.42/4.06/3.69
TURBOCHARGER make/type	NA	Garrett	NA	Garrett	NA	Garrett	NA
PRESSURIZATION (psi)	NA	NA	NA	NA	NA	NA	3.35
PRICE standard tanks (1979 $s)	52,350 to 63,235	58,995 to 69,895	60,550 to 72,100	67,450 to 79,075	67,995 to 79,845	74,995 to 86,725	103,995 to 117,030
long-range taks (1979 $s)	NA	NA	add 835	add 825	NA	NA	NA
STANDARD EQUIPMENT:							
IFR PANEL	=	=	=	=	=	=	=
VHF nav and com	II w. nav-pac	II w. nav-pac	II w. nav-pac	II w. nav-pac	II w. nav-pac	II w. nav-pac	II w. nav-pac
AUDIO PANEL	II w. nav-pac	II w. nav-pac	II w. nav-pac	II w. nav-pac	II w. nav-pac	II w. nav-pac	II w. nav-pac
MARKER	=	=	=	=	=	=	=
GLIDE SLOPE	=	=	=	=	=	=	=
TRANSPONDER	No	No	No	No	No	No	No
ENCODING ALTIMETER	=	=	=	=	=	=	=
ADF	No	No	No	No	No	No	No
DME	No	No	No	No	No	No	No
RMI	No	No	No	No	No	No	No
RNAV	No	No	No	No	No	No	No
AUTOPILOT	=	=	=	=	=	=	=

Cessna Aircraft Company twin-engine piston aircraft specifications.

CESSNA AIRCRAFT COMPANY MODEL	SKYMASTER / SKYMASTER II	TURBO SKYMASTER / TURBO SKYMASTER II	PRESSURIZED SKYMASTER / PRESSURIZED SKYMASTER II	310/II	TURBO 310/II	340/II/III	402C BUSINESSLINER	CHANCELLOR/II/III	TITAN/II/III	GOLDEN EAGLE II/III
ENGINES	CIO-360-GB (2)	C-TSIO-360-H (2)	C-TSIO-360-C (2)	C-IO-520-MB (2)	C-TSIO-520-BB (2)	C-TSIO-520-NB (2)	C-TSIO-520-VB (2)	C-GTSIO-520-L (2)	C-TSIO-520-NB (2)	C-GTSIO-520M (2)
TBO (hours)	1,500	1,400	1,400	1,700	1,400	1,400	1,400	1,400	1,400	1,200
PROPELLER	const. speed, full feathering	const. speed, full feathering	const. speed, full feathering	const. speed, full feathering	const. speed, full feathering	const. speed, full feathering	const. speed, full feathering	const. speed, full feathering	const. speed, full feathering	const. speed, full feathering
number of blades, each	2	2	2	3	3	3	3	3	3	3
SEATS	6	6	6	6	6	6	7	7	8	10
TAKEOFF ground roll (ft)	1,000	1,000	945	1,335	1,306	1,015	1,763	1,786	2,185	1,788
50' obstacle (ft)	1,675	1,675	1,500	1,700	1,662	2,175	2,195	2,323	2,595	2,387
RATE OF CLIMB (fpm)	940 (f eng 300) (r eng 320)	1,160 (1 eng 335)	1,170 (1 eng 375)	1,662 (1 eng 400)	1,700 (1 eng 390)	1,650 (1 eng 315)	1,450 (1 eng 301)	1,940 (1 eng 350)	1,580 (1 eng 290)	1,575 (1 eng 230)
Vx (knots)	81	75	69	85	81	86	84	88	88	82
Vxse (knots)	80	81	80	95	98	95	95	105	100	98
Vy (knots)	99	93	95	107	105	108	109	111	108	102
Vyse (knots)	90	87	89	106	106	100	104	111	108	102
Vmc (knots)	NA	NA	NA	80	80	82	80	82	79	78
SERVICE CEILING (ft)	16,300 (f eng 6,900) (r eng 7,100)	20,000 (1 eng 16,500)	20,000 (1 eng 18,700)	19,750 (1 eng 7,400)	27,400 (1 eng 17,200)	29,800 (1 eng 15,800)	26,900 (1 eng 14,800)	30,200 (1 eng 14,900)	31,350 (1 eng 19,850)	26,000 (1 eng 10,100)
MAX SPEED (knots)	172 (sea level)	207 (at FL 200)	211 (at FL 200)	207 (sea level)	237 (at 16,000 ft)	244 (at FL 200)	231 (at 16,000 ft)	258 (FL 200)	239 (FL 200)	232 (16,000 ft)
CRUISE SPEED at 75% (knots)	169 (at 5,500 ft)	200 (at FL 200)	205 (at FL 200)	195 (at 7,500 ft)	220 (at FL 200)	229 (at FL 245)	213 (at FL 200)	241 (FL 250)	224 (FL 245)	217 (FL 200)
at 55% (knots)	149 (at 10,000 ft)	149 (at FL 200)	179 (at FL 200)	182 (at 15,000 ft)	191 (at FL 200)	NA	179 (at FL 200)	NA	NA	NA
RANGE at 75% std tanks (nm)	545 (at 5,500 ft)	520 (at FL 200)	985 (at FL 200)	494 (at 7,500)	530 (at FL 200)	479 (at FL 245)	419 (FL 200)	955 (FL 250)	1,147 (FL 245)	1,406 (at FL 200)
at 55% std tanks (nm)	990 (at 5,500 ft)	975 (at FL 200)	1,155 (at FL 200)	884 / 1,132	1,240	899 / 1,168	711 / 1,029	1,113 / 1,271	NA	1,525 (FL 200)
long-range tanks (nm)	1,235 (at 10,000 ft)	1,125 (at FL 200)	NA	616 / 1,152 / 1,511	NA	531 / 1,036 / 1,372	484 / 836 / 234	1,287 / 1,487	1,286	1,818 (10,000 ft) / 1,836 (FL 200)
FUEL FLOW per eng. at 75% (pph)	80	87	88.8	114.5	188	128.8	NA	119	NA	114.4
at 55% (pph)	60	60	63.4	72.8	149	102.4	94.6	109.6	119	98.4
STALL CLEAN (knots)	52.8	52.8	70.62	70.62	78.70	94.6	78.68	83.74	96.8	83.70
LANDING ground roll (ft)	700	700	795	640	640	770	770	720	82.72	1,100
50' obstacle (ft)	1,650	1,650	1,675	1,700	1,790	79.71	1,850	2,293	2,393	2,130
ACCELERATE-STOP DISTANCE (ft)	2,860	3,220	3,270	3,740	3,350	1,850	3,860	3,350	4,390	8,150
RAMP WEIGHT (lbs)	4,648	4,652	4,724	5,015	5,535	5,630	6,025	3,500	4,780	8,400
ZERO FUEL WEIGHT (lbs)	NA	NA	NA	5,500 / 5,400	5,500 / 5,400	5,990 / 5,990	6,515	7,500	6,515	8,100
USEFUL LOAD (lbs) standard/II/III	4,630 / 4,400	4,630 / 4,400	4,700 / 4,465	2,082 / 1,932	2,062 / 1,812	2,116 / 1,882 / 1,769	6,850 / 6,850	7,450 / 7,200	6,750 / 6,750	8,400 / 8,100
FUEL standard tanks (lbs)	1,448 / 1,705	1,731 / 1,630	1,665 / 1,553	600	600	978 / 1,218	2,808 / 2,549	2,922 / 2,737 / 2,521	2,431 / 2,262 / 2,021	3,778 to 3,400
long-range tanks (lbs)	528	528	688	978 / 1,218	978 / 1,218	NA	900 / 1,224	1,572	1,224	2,064
MAX PAYLOAD (lbs)	888	888	NA	NA	NA	1,950	900 / 1,224	1,840	2,860	NA
PAYLOAD w. full fuel (lbs) std/II/III	1,385	1,385	1,385	1,970	1,970	1,516 / 1,282 / 1,189	2,208 to 1,949	1,686 / 1,501 / 1,285	1,207 / 1,038 / 797	3,200
FUEL w. max payload (lbs) std/II/III	1,320 / 1,177	1,245 / 1,102	777 / 665	1,582 / 1,332	1,462 / 1,212	1,119 to 571	1,518 to 1,325	1,908 to 1,325	NA	1,714 to 1,336
long-range tanks	950 / 817	865 / 742	NA	1,185 / 949	1,094 / 594	166 (not possible)	110 (not possible)	2,324 (not possible)	NA	NA
WING LOADING (lbs per sq ft)	463 / 320	388 / 245	280 / 168	212 (not possible)	92 (not possible)	32.55	30.34	34.7	29.89	573 to 200
LENGTH/HEIGHT/SPAN external (ft)	22.9	22.9	23.2	30.73	30.73	184	225.8	215	225.8	34.71
LENGTH/HEIGHT/WIDTH cabin (ft)	29.75/9.22/38.22	29.75/9.22/38.22	29.75/9.22/38.22	31.98/10.71/36.95	31.98/10.71/36.95	34.33/12.55/38.02	36.38/11.45/44.12	36.38/11.45/44.12	36.38/11.45/44.12	242
TURBOCHARGER make/type	9.92/4.29/3.65	9.92/4.29/3.65	9.42/1.23/3.65	12.64/4.17/4.04	12.64/4.17/4.04	12.67/4.08/4.38	15.83/4.28/4.67	15.84/4.25/4.63	15.83/4.25/4.63	39.53/13.26/45.33
PRESSURIZATION (psi)	NA	Garret	Garret	Garret	Garret	Garret	Garret	Garret	Garret	18.75/4.29/4.67
PRICE (1979 $s) std/II/III	NA	NA	3.35	NA	NA	4.2	NA	5	5	Garret
	94,950/110,600	109,950/125,600	143,950/161,750	133,490/158,990	154,990/180,490	213,990 to 272,490	212,990 to 263,990	334,990 to 422,990	277,990 to 362,740	5
										283,990 to 345,490

Charlie—In aviation radio phraseology the term used for the letter C.

Chart—An aeronautical or weather map.

Chase aircraft—An aircraft flown close to a second aircraft, usually for the purpose of observing performance during testing.

Chasing the needle—An expression used to imply sloppy flying.

Check list—Any list of items or procedures designed to guard against failures of the human memory. Using check lists is especially important during preflight, before takeoff and prior to landing.

Checkout—The training and flight test a pilot should undergo in order to become familiar with a particular aircraft.

Cessna Aircraft Company turboprop aircraft specifications.

CESSNA AIRCRAFT COMPANY MODEL:	CONQUEST
ENGINES	GA/TPE 331-8-401S (2)
shp (each)	635.5 (flat rated)
TBO (hours)	3,000
SEATS (crew + passengers)	2 + 7 to 9
TAKEOFF ground roll (ft)	1,785
50' obstacle (ft)	2,465
RATE OF CLIMB (fpm)	2,435 (1 eng. 715)
Vxse (knots)	110
Vyse (knots)	120
Vmca (knots)	92
ACCELERATE-STOP DISTANCE (ft)	3,750
SERVICE CEILING (ft)	37,000 (1 eng. 21,380)
MAX SPEED (knots)	295 (16,000 ft)
BEST CRUISE SPEED (knots)	293 (24,000 ft)
LONG-RANGE CRUISE SPEED (knots)	257 (33,000 ft)
RANGE w. res. best speed (nm)	2,070 (33,000 ft)
long range (nm)	2,196 (33,000 ft)
FUEL FLOW per eng. best speed (pph)	215.9
economy (pph)	184
STALL clean/dirty	90/76
LANDING ground roll (ft)	1,095
50' obstacle (ft)	1,875
RAMP WEIGHT (lbs)	9,925
ZERO FUEL WEIGHT (lbs)	8,100
TAKEOFF WEIGHT (lbs)	9,850
LANDING WEIGHT (lbs)	9,360
USEFUL LOAD (lbs)	4,336
FUEL CAPACITY (lbs)	3,182.5
MAX PAYLOAD (lbs)	3,370
PAYLOAD w. full fuel (lbs)	953.5
FUEL w. max payload (lbs)	966
WING LOADING (lbs per sq ft)	38.8
WING AREA (sq ft)	253.6
LENGTH/HEIGHT/SPAN external (ft)	39.2/13.14/49.33
LENGTH/HEIGHT/WIDTH cabin (ft)	18.75/4.25/4.63
PRESSURIZATION (psi)	6.3
PRICE (1979 $s)	995,000

Cessna Aircraft Company jet aircraft specifications.

CESSNA AIRCRAFT COMPANY MODEL:	CITATION I/ISP	CITATION II	CITATION III
ENGINES	PW/JT15D (2)	PW/JT15D-4 (2)	GA/TFE 731 (2)
THRUST (pounds, each)	2,200	2,500	3,650
TBO (hours)	3,000	2,400	NA
SEATS (crew + passengers)	2 + 8	2 + 10	2 + 10
BALANCED FIELD LENGTH, ISA (ft)	2,930	2,990	4,355
ISA + 20 (ft)	3,800	4,270	5,930
RATE OF CLIMB (fpm)	2,680 (1 eng. 800)	3,370 (1 eng. 1,055)	NA
Vr (knots)	99	107	NA
V₂ (knots)	108	114	NA
SERVICE CEILING (ft)	41,000 (1 eng. 21,000)	43,000 (1 eng. 25,200)	51,000 (1 eng. 29,500)
Mmo	.70	.70	.81
ALTITUDE CHANGEOVER (FL)	280	280	282
Vmo (knots)	260	277	320
CRUISE hi speed (knots)	347 (FL 370)	371 (FL 370)	452 (FL 430)
economy (knots)	320 (FL 410)	321 (FL 430)	377 (FL 430)
RANGE hi speed, max fuel (nm)	1,420	1,910	2,861
max payload (nm)	1,220	1,500	2,435
economy, max fuel	1,451	2,040	3,035
max payload (nm)	1,260	1,580	2,593
FUEL FLOW per eng. hi speed (pph)	416	495	542
economy (pph)	324	312	438
RAMP WEIGHT (lbs)	12,000	13,500	18,500
ZERO FUEL WEIGHT (lbs)	9,500	11,000	13,000
TAKEOFF WEIGHT (lbs)	11,850	13,300	18,300
LANDING WEIGHT (lbs)	11,350	12,700	15,700
USEFUL LOAD (lbs)	4,840	5,711	8,700
FUEL CAPACITY (lbs)	3,780	4,971	7,859
MAX PAYLOAD (lbs)	1,200	1,400	1,600
FUEL w. max payload (lbs)	3,640	4,311	7,100
PAYLOAD w. full fuel (lbs)	1,060	740	841
WING LOADING (lbs per sq ft)	42.5	41.2	58.7
WING AREA (sq ft)	278.5	322.9	312
LENGTH/HEIGHT/SPAN external (ft)	43.5/14.3/47.1	47.2/14.8/51.6	55.45/17/53.3
LENGTH/HEIGHT/WIDTH cabin (ft)	12.6/4.3/4.9	16/4.8/4.9	23/5.8/5.65
PRESSURIZATION (psi)	8.5	8.7	9.4
EXTERIOR NOISE (EPNdB)	79/88	80.1/90.5	NA (below FAR 36)
PRICE (1981 $s)	1,345,000	1,945,000	3,695,000
STANDARD EQUIPMENT:			
VHF nav & com	Yes	Yes	Yes
MARKER; GLIDE SLOPE; AUDIO PANEL	Yes	Yes	Yes
TRANSPONDER; ENCODING ALTIMETER	Yes	Yes	Yes
ADF; DME; RMI; HSI	Yes	Yes	Yes
AUTOPILOT; FLIGHT DIRECTOR	Yes	Yes	Yes
RNAV	No	Opt	Yes
VNAV	No	Opt	Yes
WEATHER RADAR	Yes	Yes	Yes
DE-ICING	Yes	Yes	Yes
GALLEY; TOILET	Yes	Yes	Yes
AIRBORNE TELEPHONE	No	No	Yes

Check pilot—The pilot who checks out another pilot.

Check point—A geographical point or prominent landmark the location of which can be determined by reference to a chart and identified either visually or by radio.

Cheetah—A single-engine fixed-gear piston aircraft manufactured by Gulfstream-American Corporation.

Cherokee—A family of low- and high-performance aircraft produced by Piper Aircraft Corporation.

Cheyenne—A family of turboprop aircraft produced by Piper Aircraft Corporation.

CHG—Change.

Chieftain—A piston aircraft of the cabin-twin class produced by Piper Aircraft Corporation.

Chord—An imaginary line running from the leading edge to the trailing edge of the wing.

CHT—Cylinder-head-temperature gauge.

CIG—Ceiling.

Canadair Challenger 600 (courtesy of Canadair).

Canadair Ltd. jet aircraft specifications.

CANADAIR, LTD.	CHALLENGER 600
ENGINES	L/ALF-502-L
THRUST (pounds)	7,500 each
TBO	4,000 (core), 6,000 (compressor) 10,000 (fan)
SEATS	2 crew + 11 – 19
BALANCED FIELD LENGTH, ISA (ft)	4,300
ISA + 20 (ft)	5,800
RATE OF CLIMB (fpm)	6,000 (1 eng. 1,555)
Vr (knots)	125
V2 (knots)	130
SERVICE CEILING (ft)	49,000 (1 eng. 28,000 at 32,000 lbs)
Mmo	0.88
ALTITUDE CHANGEOVER (FL)	250
Vmo	375
CRUISE high speed FL 410 (knots)	476
long range FL 410 (knots)	460
RANGE hi speed w. max fuel & res. (nm)	3,700
w. max pld. & res. (nm)	1,900 (15 passengers)
economy, w. max fuel & res. (nm)	4,250
w. max pld. & res. (nm)	3,400 (15 passengers)
FUEL FLOW per engine, hi speed (pph)	950
economy, (pph)	700
RAMP WEIGHT (lbs)	32,650
ZERO FUEL WEIGHT (lbs)	24,600
TAKEOFF WEIGHT (lbs)	32,500
LANDING WEIGHT (lbs)	31,000
USEFUL LOAD (lbs)	15,400
FUEL CAPACITY (lbs)	14,610
MAX PAYLOAD (lbs)	7,500
FUEL w. max payload (lbs)	7,900
PAYLOAD w. full fuel (lbs)	940
WING AREA (sq ft)	450
LENGTH/HEIGHT/SPAN external (ft)	68.5/20.7/61.8
cabin (ft)	25.1/6.1/8.2
PRESSURIZATION (psi)	9.45
EXTERIOR NOISE (EPNdB)	78 (takeoff), 87 (sideline), 90 (approach)
PRICE (U.S. $s)	7.000.000

STANDARD EQUIPMENT: VHF NAV & COM/MARKER/GLIDE SLOPE/AUDIO
PANEL/TRANSPONDER/ENCODINGDAR ALTIMETER/WEATHER RADAR/
THRUST REVERSER/DE-ICING

Cessna Chancellor (courtesy of Cessna).

Circle to land—A maneuver executed by the pilot to line himself up on final approach to the active runway, when a straight-in landing from a given instrument approach is not possible. The maneuver requires prior ATC authorization and is permissible only after the pilot has established visual contact with the airport.

Circle to runway (number)—A clearance expression used by ATC to inform the pilot that he must circle to land because the active runway is not aligned with the particular instrument-landing procedure being flown.

Circling approach—See circle to land.

Circling minimums—Circling minimums are published on instrument approach charts and provide adequate obstruction clearance and pilots must not descend below the circling MDA until visual contact has been established with the airport and the aircraft is in a position to make the final descent for landing.

Gulfstream-American Cheetah (courtesy of Gulfstream-American).

Piper Cherokee Six (courtesy of Piper).

Circuit breaker—A means of interrupting the flow of electrical current when an overload occurs. It can be reset by pushing a button. Its function is identical to that of a fuse.

Cirrocumulus—A high cumuliform cloud occurring in bright puffs. When joined in a mass they have a rippling appearance, frequently referred to as "mackerel sky".

Cirrostratus—Cirrus cloud in a solid or slightly broken layer, often topped by individual cirrus clouds.

Cirrus—A high stratiform cloud which occurs in bright filmy streaks or whisps through which the sun is easily visible. It consists of ice crystals.

Citabria—One of the Champion family of sport aircraft manufactured by Bellanca Aircraft Corporation.

Citation—A family of executive jet aircraft manufactured by Cessna Aircraft Company.

Civil aviation—All non-military aviation.

Piper Cheyenne II (courtesy of Piper).

Piper Chieftain (courtesy of Piper).

CL—Control tower.

CLC—Course line computer; a component of RNAV systems.

CLD—Cloud.

Clean—Refers to aerodynamically clean, meaning with landing gear, flaps and spoilers retracted.

Clear—Sky conditions in which clouds cover less than 10 percent of the sky.

Clear air turbulence—(CAT)—Turbulence occurring in clear air and not associated with any cloud formation. Occurring most frequently in the vicinity of the jet stream.

Clearance—An authorization by ATC for an aircraft to proceed under specified conditions within controlled airspace. Clearances are given based on known traffic in the affected area.

Clearance delivery—At busy airports a separate ATC function with its own frequency, solely for the purpose of issuing clearances to departing IFR flights.

Clearance limits—A point, determined either by time or navigation aids, at which the pilot must have further clearance from ATC in order to proceed.

Clearance shorthand—Shorthand used by pilots when copying an ATC clearance. Use of this shorthand is recommended but not mandatory.

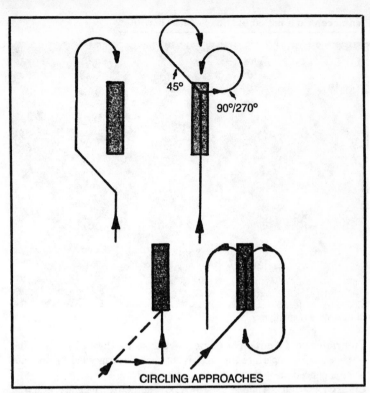

CIRCLING APPROACHES

Circling approaches (courtesy of FAA).

Cirrus clouds occur in whisps.

Cirrus in the evening sky.

Clearance void if not off by (time)—A phrase used by ATC to inform the pilot that his departure clearance is automatically cancelled if he has not commenced his flight by the time given.

Cleared as filed—A phrase used by ATC to authorize a pilot to proceed on an IFR flight in accordance with the way in which he filed his IFR flight plan.

Cleared for approach—An authorization by ATC for the pilot to execute any type of published instrument approach of his choice.

Cleared for (type of) approach—An authorization by ATC for the pilot to commence the instrument approach specified in the clearance.

Cleared for takeoff—An ATC authorization for an aircraft to depart. It is frequently issued as "Cleared for immediate takeoff", when other traffic is on final approach.

Cleared for the option—An ATC authorization for an aircraft to execute a special maneuver such as a touch-and-go, stop-and-go, missed approach, etc.

Cleared through—An ATC authorization for an aircraft to make intermediate stops at specified airports without having to refile his flight plan.

Cleared to land—An ATC authorization for an aircraft to land.

Clear ice—Ice which forms in smooth transparent layers from the gradual freezing of supercooled water. It is most prevalent on the smooth surfaces of an aircraft and tends to form most frequently when flying through freezing drizzle. Also called *glaze*.

Clear of traffic—A phrase used by ATC to inform a pilot that previously issued traffic is of no further consequence.

Citabria (courtesy of Bellanca).

Climb—The portion of a flight during which the aircraft ascends from the ground to its cruising level; or any other time when the aircraft changes to a higher altitude or flight level.

Climb-and-descent corridor—A narrow portion of airspace, usually established near military airports, where high-speed jet aircraft can climb or descend at speeds above 250 knots.

Climbing turn—A turn, usually to a predetermined heading, made while climbing. It requires added power to maintain constant airspeed and rate of climb.

Cessna Citation (courtesy of Cessna).

Climb to VFR—An AC authorization, usually issued in conjunction with a Special VFR clearance, for an aircraft to climb to VFR conditions within a control zone when the only weather factor is restricted visibility. It requires that the aircraft remain clear of clouds during the climb.

Climbout—Flight between takeoff and cruising altitude.

Clock—A clock with a sweep-second hand is a requirement for IFR flight. Recently, a wide variety of digital clocks and timers have appeared on the market, most designed to simplify the timing problems associated with non-precision instrument approaches.

Closed runway—A runway which, for one reason or another, is unusable for aircraft operation. Permanently closed runways must be marked by a large X.

Closed traffic—Continuous activity involving takeoffs and landings during which the aircraft does not leave the traffic pattern.

Cloud—A visible mass of small water droplets condensed from the water vapor in the atmosphere, or, at higher altitudes, ice crystals. Cloud formation requires the presence of condensation nuclei and a drop in air temperature below the dew point at the altitude at which the formation takes place.

Cloud bank—A well-defined mass of clouds seen in the distance, covering a considerable portion of the horizon or sky.

Cloud cover—Sky cover.

CLR—Clear.

CLRNC—Clearance.

CLSD—Closed.

Clutter—A term used to describe radar returns caused by precipitation, terrain, chaff, large numbers of aircraft or any other phenomenon producing an excessive number of targets in close proximity, and often making it difficult or impossible for ATC to provide effective radar service.

CNTR—Center.

CNTRL—Central.

Cockpit—The portion of an aircraft fuselage occupied by the flight crew.

Codes—The number assigned by ATC to a transponder-equipped aircraft. VFR aircraft not in contact with ATC always squawk 1200. IFR aircraft and VFR aircraft under ATC control will be asked by ATC to squawk a given code number.

Col—A narrow neck between two highs or two lows, of the same pressure as the centers.

Cold air mass—A mass of unstable air, colder than the surface over which it is moving. When warmed it results in convection currents and, in turn, clouds with vertical development. Visibility is usually good.

Cold front—A front formed by a mass of cold high-pressure air moving under warm air and replacing it. It is usually associated with turbulent cumulonimbus clouds and line squalls. More often than not it doesn't last long because the front is shallow. It is indicated on a weather chart by a

Cold air mass.

line with pointed triangular marks in the direction in which the front is moving.

Colemill Enterprises, Inc.—A major modifier of twin-engine aircraft. (P.O. Box 60627, Nashville, TN 37206.)

Collective—One of the primary flight controls on a helicopter.

Collins Avionics Division—A subsidiary of Rockwell International. A major manufacturer of high-quality avionics. (400 Collins Road NE, Cedar Rapids, IA 52406.)

Collision avoidance system—Cockpit instrumentation capable of warning the pilot of the proximity of other aircraft, especially those which could pose a threat if both aircraft continue on course.

Com—Communications.

Comanche—A high-performance single- and twin-engine aircraft manufactured by Piper Aircraft Corporation. Comanches are no longer in production.

Combined station/tower—A facility at which the functions of a control tower and a flight service station are combined.

Combustion chamber—In a reciprocating engine the vacant area between the cylinder head and the highest point reached by the stroke of a piston. If is here that combustion of the fuel and air mixture occurs when ignited by the spark plug.

Comm—Communications.

Piper Comanche.

Commander—A family of single- and twin-engine aircraft manufactured by the General Aviation Division of Rockwell International. It includes high-performance singles, piston twins and turboprops.

Commercial license—A pilots license authorizing the pilot to carry passengers or freight for remuneration.

Commercial pilot—See *commercial license.*

Common route—A segment of a route between an inland navigation aid and a coastal fix.

Communications Components Corporation—A manufacturer of sophisticated avionics equipment, especially long-range navigation equipment. (3000 Airway Avenue, Costa Mesa, CA 92626.)

Compass—A device for determining the direction of flight in the horizontal plane. The magnetic compass aligns itself automatically with the magnetic north. It tends to be unreliable during turns, climbs and descents. A gyro compass must be intermittently set by the pilot to conform with the magnetic compass. (See also directional gyro.)

Compass card—A circular scale in a magnetic compass or directional gyro, showing compass headings in degrees from 0 to 359. The readout may consist of a needle moving around the scale, or the card itself revolving with relation to a lubber line.

Compass correction card—A card located in the cockpit and showing compass errors at various directions, caused by interference due to the airframe or instruments.

Rockwell Commander 700 (courtesy of Rockwell International).

Compass deviation—See compass card.

Compass error—Error in the reading of a magnetic compass induced by turns executed by the aircraft.

Compass heading—The compass heading which must be flown in order to achieve a planned true course. It is arrived at by correcting the true course for magnetic variation, compass deviation and wind-drift angle.

Compass locator—A low or medium frequency low-power radio beacon installed in conjunction with an ILS at the outer or middle marker. It is not usable for navigation at distances greater than 15 nm.

Compass rose—See compass card.

Composite flight plan—A flight plan which combines VFR and IFR legs during one flight.

Rockwell Commander Medalist (courtesy of Rockwell International).

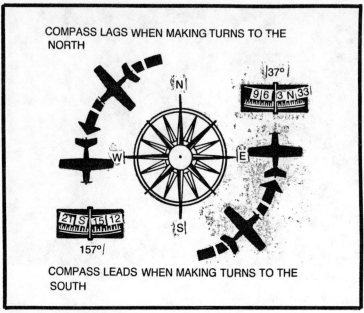

Northerly and southerly turn errors (courtesy of FAA).

Compression—The degree of compression of the fuel and air mixture in reciprocating and turbine engines.

Compressor—The section of a turbine engine which compresses the air to the desired degree.

Compressor blades—Small metal blades attached to the compressor in a turbine engine.

Compulsory reporting point—Any position along an airway at which a pilot must report his position to ATC when not in radar contact. Compulsory reporting points are shown on radio-facility charts as solid black triangles.

Computer—An electronic instrument capable of extremely fast calculation and usually equipped with a memory which stores repeatedly used information. Two types of computers, analog and digital, are used in aviation, both in cockpit instrumentation and by ATC. Small programmable electronic pocket computers manufactured by several companies are capable of performing all manner of aviation-related computing chores and, in some instances, can be used to program sophisticated RNAV equipment or to cause check lists or other information to be displayed on the airborne weather radar display.

COMSND—Commissioned.

COMSNG—Commissioning.

Com transceiver—A combination radio receiver and transmitter used for two-way radio communication. It may utilize VHF or HF frequencies.

COND—Condition.

Condensation—The process by which water vapor changes into liquid, usually because of a decrease in temperature. It is the opposite of evaporation.

Condensation level—The altitude at which a rising column of air reaches the condensation point and clouds begin to form.

Condensation nuclei—Impurities in the atmosphere, such as dust or sand, around which water vapor condenses to form precipitation or clouds.

Condensation trail—A visible trail of condensed water vapor or ice crystals left behind an aircraft. Also called *contrail* or *vapor trail*.

Conduction—The transfer of heat energy by contact of a cool region with a warm region.

Cone of silence—The airspace above a LF/MF nav aid in which no range signals can be heard because of the directional nature of the signals. It is usually co-located with a marker beacon.

Cones—Cone shaped nerve ends in the center of the retina of the human eye. Cones are capable of distinguishing colors, but are ineffective under conditions of very low light levels. (See night vision.)

CONFIG—Configuration.

Conflict alert—An advisory by ATC with reference to conflicting traffic, usually in the form of: "Advise you turn right/left…climb/descend…etc."

Connecting rod—A metal bar which converts the back-and-forth motion of a piston into the rotating motion of the crankshaft in a reciprocating engine.

Conquest—An executive turboprop twin-engine aircraft manufactured by Cessna Aircraft Company.

Consolan—A low-frequency long-distance nav aid useful in transoceanic navigation.

Constant pressure chart—A chart showing the position of a line of constant pressure with changes in altitude in the upper air. It is plotted every 12 hours from measurements made by radiosondes.

Com transceiver (courtesy of King Radio).

Com transceivers comparison chart (continued on page 93).

MANUFACTURER	MODEL	PRICE uninstalled	Volts input	Frequencies number	Frequencies range MHz	Frequencies spacing kHz	storage	Squelch automatic	Squelch manual	Mount panel	Mount remote	Portable unit avail	Transmitter output (watts)	Units	Weight lbs
MENTOR	TR-12	432	14	1-10	118-135.95	50							2	1	2
GENAVE	ALPHA/10	650	14	10	118-135.95	50							4	1	4
GENAVE	ALPHA/100	750	14 28*	100	118-127.9	100							2-3	1	4
RADAIR	360	850	14	360	118-135.95	50							6	1	2.8
MENTOR	M-360	880	14 28*	360	118-135.95	50							10	1	3
EDO-AIRE/ FAIRFIELD	RT-551	945	14 28*	360	118-135.95	50							6	1	2.7
COLLINS	VHF-250	1.045	14	720	118-135.975	25							10	1	3.3
KING	KY 92	1.045	14 28*	720	118-135.875	25							7	1	2.8
EDO-AIRE/ FAIRFIELD	RT-661	1.045	28*	360	118-135.975	50							6	1	2.7
EDO-AIRE/ FAIRFIELD	RT-551A	1.045	14 28*	720	118-135.975	25							6	1	2.7
EDO-AIRE/ FAIRFIELD	RT-661 A	1.145	14 28*	720	118-135.975	25							6	1	2.7
GENAVE	ALPHA/720	1.240	14 28*	720	118-135.975	25							4	1	4
KING	KY 195 B	1.315	14 28*	720	118-135.975	25	1						7	1	6.3
NARCO	COM 120	1.345	14 28*	720	118-135.975	25							12	1	3.5
COLLINS	VHF-251	1.375	14	720	118-135.975	25	1						10	1	3.4
RADAIR	SKY-515A	1.600	14 28*	380	117-135.95	50							5	1	3.4
KING	KY 196	1.755	28	720	118-135.975	25	1						16	1	3.2
		1.755	14	720	118-135.975	25	1						10	1	3.2
WULFSBERG	WT-200	2.328	28	720	118-135.975	25							20	1	5.6
BECKER FLUGFUNKWERK	AR-2009	2.360	14 28*	720	118-135.975	25							8	1	2.9
NARCO	COM 200	2.800	14 28*	720	118-135.975	25	1						20	1	3.5
KING	KTR 905	2.830	28	720	118-135.975	25							20	1	4.4
ELECTRONIQUE AEROSPATIOLE	TR 800R	2.850	28	720	118-135.975	25							20	1	5.5
ELECTRONIQUE AEROSPATIALE	TR 800 RM	2.900	28	1360	116-149.975	25							20	1	5.5
BECKER FLUGFUNKWERK	AR 2010	2.950	14 28*	720	118-135.975	25							8	1	2
WULFSBERG	WT-2000	3.328	28	720	118-135.975	25							20	1	8.9
ELECTRONIQUE AEROSPATIALE	TR800	3.400	28	720	118-135.975	25	4						20	1	6.4
ELECTRONIQUE AEROSPATIALE	TR 800A	3.400	28	1040	118-143.975	25	4						20	1	5.4
KING	KTR 9100 A	3.427	28	720	118-135.975	25							25	1	13

(Left margin grouping labels, repeated vertically: VHF TRANSCEIVERS)

Com transceivers comparison chart (continued from page 92).

MANUFACTURER	MODEL	PRICE uninstalled	Volts input	Frequencies number	range MHz	spacing kHz	storage	Squelch automatic	manual	Mount panel	remote	Portable unit avail	Transmitter output (watts)	Units	Weight lbs
COLLINS	VHF-20A	3.660	28	720	118-135.975	25							20	1	5.8
COLLINS	618M-3	3.904	27.5	720	118-135.975	25							25	1	10
BECKER FLUGFUNKWERK	AR 2011	3.950	14 28*	720	118-135.975	25							10	1	2
BENDIX AVIONICS	RTA-43A	3.996	28	720	118-135.975	25							25	1	11.8
ELECTRONIQUE AEROSPATIALE	TTR 730	4.000	28	720	118-135.975	25							25	1	12.8
MARCONI AVIONICS	AD 120	4.489	28	720	118-135.975	25							20	1	5
COLLINS	G18M-3A	4.568	27.5	1440	116-151.975	25							25	1	10
ARC	RT-1038A	4.880	28	720	118-135.975	25 50	3						20	2	7.6
COLLINS	VHF-20 B	4.995	28	1360	118-151.975	25								1	5.8
ELECTRONIQUE AEROSPATIALE	TTR 730M	5.000	28	1360	116-149.975	25							25	1	12.8

Constant-speed propeller—A controllable-pitch propeller which maintains a constant rpm by automatically changing the blade angle in relation to engine output.

CONSTR—Construction.

CONSTRD—Constructed.

CONT—Continue, continuous.

Contact—A verbal warning shouted by the pilot, indicating that the engine is about to be started. An alternate term used is "Clear!" or "Clear prop!"

Contact approach—A visual approach made by an aircraft on an IFR flight plan when operating with at least one mile visibility and when authorized by ATC. This approach must initially be requested by the pilot.

Contact conditions—Weather conditions under which a pilot can navigate by reference to the ground.

Contact flying—Flying by reference to the ground under visual flight rules (VFR).

Conterminous U.S.—The 48 adjoining states and the District of Columbia.

Contiguous U.S.—See Conterminous U.S.

Continental control area—The airspace above 14,500 feet msl and at least 1,500 feet agl above the contiguous U.S. and most of Alaska. Prohibited and resticted areas are not included in the continental control area.

Continental U.S.—The 49 states located on the North American continent plus the District of Columbia.

Cessna Conquest (courtesy of Cessna).

Continuous transcribed weather broadcast—A continuous transmission of transcribed weather information and pertinent PIREPs by LF/MF stations and selected FSSs.

Contour lines—Lines on aeronautical charts which link points of the same elevation and thus indicate ground relief.

Contrail—Condensation trail.

Control area—See controlled airspace.

Controllable-pitch propeller—A propeller the blade angle of which may be changed by the pilot in flight in order to obtain the best or most economical performance.

Controlled airport—An airport at which all arriving and departing traffic and all traffic passing through the airport traffic area is governed by ATC.

Controlled airspace—All airspace in which all IFR traffic is subject to ATC. Also including all positive-control airspace in which all traffic is subject to ATC.

Controller—An employee of the FAA authorized to provide air-traffic-control service at en-route and terminal ATC facilities. Not included is FSS personnel, known as flight-service specialists.

Controls—Any and all devices used by the pilot in the process of operating an airplane.

Control sector—An area of airspace within horizontal and vertical limits over which a controller or group of controllers has jurisdiction. Usually a portion of the area controlled by a given ARTCC or approach/departure control facility.

Control stick—Control wheel or the yoke in an aircraft. The term is applicable whether it is a wheel or actually a stick as in older and some current sport aircraft.

Control surface—Any movable airfoil such as aileron, rudder, elevator, trim-tab, which can be operated by the pilot in order to achieve a desired reaction.

Control tower—Airport traffic control tower, the ATC facility at a controlled airport.

Control wheel—Control stick; yoke.

Control zone—A more or less circular area around a controlled airport, 10 miles in diameter including extensions necessary for instrument ap-

Airport traffic control tower.

proaches, which is under ATC control. Control zones extend from the surface upward to the base of the continental control area or, where not underlying the continental control area, with no upper limit. They are shown on aeronautical charts by a broken blue line.

Convection—The transfer of heat through the atmosphere by the motion of vertical columns of air, usually resulting from uneven heating of the ground and producing turbulent conditions.

Convection current—A vertical current of air, commonly referred to as up or downdraft.

Convection fog—Fog, usually in the vicinity of large bodies of water, resulting from the air currents produced by the difference in heat absorption of land and water.

Conventional gear—Tail-wheel gear. Such airplanes are commonly known as taildraggers.

Conversion charts—Conversion tables.

Conversion tables—Tables of charts showing the comparative values of different means of identifying weights and measures:

Centigrade/Fahrenheit Conversion Table

C.	F./C.	F.	C.	F./C	F.
-62.2	-8.0	-112	68.3	155	311
-56.7	-70	-94	71.1	160	320
-51.1	-60	-76	73.9	165	329
-45.6	-50	-58	76.7	170	338
-40	-40	-40	79.4	175	347
-34.4	-30	-22	82.2	180	356
-31.7	-25	-13	85	185	365
-28.9	-20	-4	87.8	190	374
-26.1	-15	+5	90.6	195	383
-23.3	-10	14	93.3	200	392
-20.6	-5	23	96.1	205	401
-17.8	0	32	98.9	210	410
-15	5	41	101.7	215	419
-12.2	10	50	104.4	220	428
-9.4	15	59	107.2	225	437
-6.7	20	68	110	230	446
10	50	122	112.8	235	455
12.8	55	131	115.6	240	464
15.6	60	140	118.3	245	473
18.3	65	149	121.1	250	482
21.1	70	158	126.7	260	500
			132.2	270	518
			137.8	280	536
			143.3	290	554

Centigrade/Fahrenheit Conversion Table

C.	F./C.	F.	C.	F./C	F.
148.9	300	572	165.6	330	626
154.4	310	590	171.1	340	644
160	320	608	176.7	350	662
40.6	75	167	182.2	360	680
43.3	80	176	187.8	370	698
46.1	85	185	193.3	380	716
48.9	90	194	198.9	390	734
51.7	95	203	204.4	400	752
23.9	100	212	210	410	770
26.7	105	221	215.6	420	788
29.4	110	230	221.1	430	806
32.2	115	239	226.7	440	824
35	120	248	232.2	450	842
37.8	125	257	237.8	460	860
54.4	130	266	243.3	470	878
57.2	135	275	248.9	480	896
60	140	284	254.4	490	914
62.8	145	293	260	500	932
65.6	150	302	265.6	510	950

Velocity Conversion Table

knots	mph/knots	mph	knots	mph/knots	mph
4	5	6	69	80	92
9	10	12	74	85	93
13	15	17	78	90	104
17	20	23	82	95	110
22	25	29	87	100	115
26	30	35	91	105	121
30	35	40	95	110	127
35	40	46	100	115	132
39	45	52	104	120	138
43	50	58	108	125	144
48	55	63	113	130	150
52	60	69	117	135	155
56	65	75	122	140	161
61	70	81	126	145	167
65	75	86	130	150	173

To convert mph into knots, divide by 1.15. To convert knots into mph, multiply by 1.15.

Miscellaneous Conversion Factors

multiply	by	to obtain
Acres	43,560	square feet
	4,047	square meters
	1.562×10^{-3}	square miles
Atmospheres	76	cm of mercury
	29.921	inches of mercury
	33.899	feet of water
	10,332	kilogram per sq. meter
	14,696	pounds per sq. inch
	2,116.2	pounds per sq. foot
	1.0133	bars
Bars	75.01	cm of mercury
	14.5	pounds per sq. inch
BTU	778.2	foot pounds
	$.3930 \times 10^{-3}$	horsepower hour
	$.2930 \times 10^{-3}$	kilowatt hour
	.2520	kilogram calorie
	107.6	kilogram meters
	1055	joules
Centimeters (cm)	.3937	inches
	.03281	feet
cm of mercury	5.3524	inches of water
	.44603	feet of water
	.19337	pounds per sq. inch
	27.845	pounds per sq. foot
	135.95	kilogram per sq. meter
Cubic centimeters	10^{-3}	liters
	.06102	cubic inches
Cubic feet	28,317	cubic centimeters
	1,728	cubic inches
	.02831	cubic meters
	7.4805	gallons
	28.316	liters
Cubic feet per min.	.4717	liters per second
	.02832	cubic meters per min.
Cubic feet of water	62.428	pounds
Cubic inches	16.387	cubic centimeters
	.01639	liters
	4.329×10	gallons
	.01732	quarts

Miscallaneous Conversion Factors

multiply	by	to obtain
Cubic meters	61,023	cubic inches
	35,314	cubic feet
	264.17	gallons
Cubic yards	27	cubic feet
	.7646	cubic meters
	202	U.S. gallons
Degrees (arc), dynes	.01745	radians
	1.020×10^{-3}	grams
	2.248×10^{-6}	pounds
	7.233×10^{-5}	poundals
Ergs	$.947 \times 10^{-10}$	BTU
	1	dyne centimeter
	7.376×10^{-8}	foot pounds
	1.02×10^{-3}	gram centimeters
	10^{-7}	joules
	2.388×10^{-4}	kilogram calories
Feet	.3048	meters
Feet of water	.0295	atmospheres
	.43353	pounds per sq. inch
	62.378	pounds per sq. foot
	304.8	kilogram per sq. meter
	.88367	inches of mercury
	.24199	centimeters of mercury
Feet per minute	.01136	miles per hour
	.01829	kilometers per hour
	.508	centimeters per second
	.009878	knots
Feet per second	.68182	miles per hour
	1.0973	kilometers per hour
	30.48	centimeters per second
	.3048	meters per second
	.59209	knots
Foot-pounds	.13826	meter-kilogram
Foot-pounds/min.	.00003	horsepower
Foot-pounds/sec.	.00182	horsepower
Gallons (imperial)	277.4	cubic inches
	1.201	U.S. gallons
	4.546	liters
Gallons, U.S., dry	268.8	cubic inches
	.1556	cubic feet
	1.164	U.S. gallon, liquid
	4.405	liters

Conversion tables

Miscellaneous Conversion Factors

multiply	by	to obtain
Gallons, U.S., liquid	231	cubic inches
	.13368	cubic feet
	3.7853	liters
	.83268	imperial gallons
	128	liquid ounces
Kilogram-meters	7.233	foot pounds
	9.8067×10^7	ergs
Kilogram per cu m	.06243	pounds per cubic foot
	.001	grams per cubic centimeter
Kilogram per meter	.67197	pounds per foot
Kilogram per sq. m	.00142	pounds per sq. inch
	.20482	pounds per sq. foot
	.0029	inches of mercury
	.00328	feet of water
	.1	grams per sq. centimeter
Kilometers	3,280.8	feet
	.62137	miles
	.53956	nautical miles
Kilometers per hr.	.91134	feet per second
	.53955	knots
	.62137	miles per hour
	.2777	meters per second
Kilowatts	.948	BTU per second
	737.7	foot-pounds per second
	1.341	horsepower
	.2389	kilogram calories per sec.
Knots	1	nautical mile per hour
	1.6889	feet per second
	1,1516	miles per hour
	1.8532	kilometers per hour
	.51479	meters per second
Liters	1,000	cubic centimeters
	61.025	cubic inches
	.03532	cubic feet
	.26418	gallons
	.21998	imperial gallons
Meters (m)	39.37	inches
	3.2808	feet
	1.0936	yards
Meters per second	3.2808	feet per second
	2.2369	miles per hour
	3.6	kilometers per hour
	1.9451	knots
Miles (statute)	5,280	feet
	1.6093	kilometers
	.86839	nautical miles

Miscellaneous Conversion Factors

multiply	by	to obtain
Miles per hr. (mph)	1.4667	feet per second
	.44704	meters per second
	1.6093	kilometers per hour
	.86839	knots
Grams	15.432	grains
	.03527	ounces
	.0022	pounds
	1,000	milligrams
	.001	kilograms
	980.67	dynes
Gram-calories	.00397	BTU
Grams per cm	.1	kilograms per meter
	.0672	pounds per foot
	.00559	pounds per inch
Grams per cu cm	1,000	kilograms per cu cm
	62.428	pounds per cu foot
Horsepower	33,000	foot-pounds per minute
	550	foot-pounds per second
	76.040	kilogram-meters per sec.
	1.0139	metric horsepower
Horsepower, metric	75	kilogram-meters per sec.
	.98632	horsepower
Horsepower-hours	2,545.1	BTU
	1,980,000	foot-pounds
	273,745	kilogram-meters
Inches	2.54	centimeters
Inches of mercury at 0°C.	.03342	atmospheres
	13.595	inches of water
	1.1329	feet of water
	.49116	pounds per sq. inch
	70.727	pounds per sq. foot
	345.32	kilograms per sq. meter
Inches of water	.07356	inches of mercury
	.18683	centimeters of mercury
	.03613	pounds per sq. inch
	5.1981	pounds per sq. foot
	25.4	kilograms per sq. meter
Joules	$.9478 \times 10^{-3}$	BTU
	.7376	foot-pounds
	$.2388 \times 10^{-3}$	kilogram calories
	.10179	kilogram meters
	$.2777 \times 10^{-3}$	watt hours
	$.3725 \times 10^{-6}$	horsepower hours
Kilograms	2.2046	pounds
	32.274	ounces
	1,000	grams

Miscellaneous Conversion Factors

multiply	by	to obtain
	3.9685	BTU
Kilogram calories	3,087.4	foot-pounds
	426.85	kilogram-meters
Nautical miles	6080.2	feet
	.0625	pounds (avdp)
Ounces (avdp)	28.35	grams
	437.5	grains
Ounces, fluid	29.57	cubic centimeters
	1.805	cubic inches
Pounds	453.59	grams
	7,000	grains
	16	ounces
	32.174	poundals
Pounds per cu ft	16.018	kilograms per cu m
	.01602	grams per cu cm
Pounds per cu inch	1,728	pounds per cu ft
	27,680	grams per cu cm
Pounds per sq inch	2.0361	inches of mercury
	2.3066	feet of water
	.06805	atmospheres
	703.07	kilograms per sq meter
	.07031	kilograms per sq cm
Radians	57.296	degrees (arc)
Radians per sec.	57.296	degrees per second
	.15916	revolutions per second
	9.8493	revolutions per minute
Revolutions	6.2832	radians
Revolutions per min.	.10472	radians per second
Square centimeters	.155	square inches
	.00108	square feet
Square feet	929.03	square centimeters
	144	square inches
	.0929	square meters
	.111	square yards
Square inches	645.16	square millimeters
	6.4516	square centimeters
Square kilometers	.3861	square miles
Square meters	10.764	square feet
	1.196	square yards
Square miles	2.59	square kilometers
	640	acres
Square yards	.83613	square meters
Yards	.9144	meters

For density altitude conversions, see *density altitude*.

For Mach/knots conversions, see *Mach*.

For the effects of pressurization on cabin altitude, see *pressurization*.

Coordinated turn—A smooth turn accomplished by using the proper amount of aileron and rudder to prevent the airplane from either slipping or skidding. In a coordinated turn the ball in the turn-and-bank indicator remains centered.

Coordinates—Latitudes and longitudes determining a given geographical point on the surface of the earth.

Coordination fix—A fix where ATC facilities will turn the control of an IFR aircraft over from one to another.

Copilot—The second in command, occupying the right seat in the cockpit.

Copy—A term which refers to having received and understood a radio message, such as in: "Did you copy?" It has nothing to do with writing it down.

Coriolis force—A force caused by the rotation of the earth which, in the northern hemisphere deflects moving bodies to the right of their course and has the opposite effect in the southern hemisphere. It does not have an effect on their velocity and it tends to make winds move parallel to the isobars.

Corporate aviation—Aviation activity conducted by major corporations for the purpose of improving productivity. With the recent trend of locating factories and facilities away from major metropolitan areas, corporate aviation has become increasingly important. More than half of FORTUNE magazine's list of the 1,000 largest corporations operate their own aviation departments with fleets of aircraft.

Correction—In aviation-radio communication a phrase which means that an error has been made in the transmission.

Cougar—A light-light twin-engine aircraft manufactured by Gulfstream-American Corporation.

Coupler—A means of coupling an autopilot to any number and variety of nav instruments, such as a VOR, localizer, flight director, etc.

Course—The planned direction of flight in the horizontal plane.

Gulfstream-American Cougar (courtesy of Gulfstream-American).

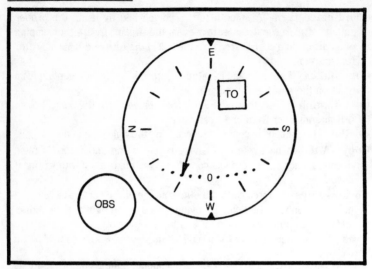

Traditional course deviation indicator needle.

Course—Any leg of an LF/MF range.

Course deviation indicator—(CDI)—The vertical needle of an omni-bearing indicator (OBI) which shows where the aircraft is in relation to the VOR radial selected on the omni-bearing selector (OBS). On course, the needle is centered. Off course, the radial is located in the direction that the needle has moved away from center.

Course line computer—(CLC)—An integral part of area navigation systems, affecting the electronic relocation of VORTAC stations.

Course selector—Omni-bearing selector (OBS).

Cowl flap—A movable door or shutter in an engine cowling designed to regulate the flow of cooling air around the engine. It may be adjusted on the ground or in the air to achieve the desired cylinder-head temperatures.

Cowling—A removable cover or housing containing the engine or any other aircraft component.

Crab—To turn partly into the wind to the right or left of course in order to compensate for wind drift.

Crab angle—Wind correction angle.

Crankshaft—A rotating shaft which, by means of the connecting rod, gives movement to the pistons in a reciprocating engine, as well as transferring piston motion to the propeller.

Critical altitude—The highest altitude at which, due to decreasing atmospheric density, an airplane can maintain its maximum allowable continuous power setting. Above this altitude, even with full throttle, the power will begin to fall off. Turbocharging reciprocating engines greatly increases the critical altitude.

Critical angle of attack—See burble point.

Critical engine—The engine in a twin-engine aircraft which, if it fails, would most adversely affect the performance and handling characteristics of the aircraft.

Cross country—Any flight other than a local flight which requires some degree of navigation, usually to an airport 25 or more miles distant from the takeoff point. A cross-country flight returning to the point of departure without intermediate landing is referred to as a *round robin*.

Cross (fix) at (altitude)—An ATC clearance requiring the pilot to cross a certain fix at a given altitude.

Cross (fix) at or above (altitude)—An ATC clearance requiring the pilot to cross a certain fix at or above a given altitude.

Crossing altitude—The minimum altitude which must be maintained when crossing a given fix or area.

Cross pointer instrument—An instrument such as the combination localizer/glide slope indicator or certain flight-director pictorial displays which require coordinating separate vertical and horizontal readings on one instrument.

Crosswind—Wind blowing at any angle across the line of flight and causing the aircraft to drift.

Crosswind component—The wind component, in knots, in 90 degrees to the direction of the runway or course. It can be figured out by using the wind-correction portion of the average flight computer.

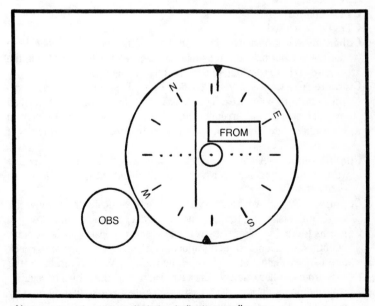

New version of a course deviation indicator needle.

Crosswind leg—A flight path at right angles to the landing runway off its upwind side.

CRS—Course.

CRT—Cathode ray tube.

Cruise—To fly at a speed a given percentage below maximum power at a constant altitude, which results in good range and economy.

Cruise—A phrase used by ATC in a clearance to indicate that the pilot may climb to the assigned altitude and may leave it at his own discretion.

Cruise climb—Climbing at an angle less than that required for the best rate of climb. Though it takes longer to arrive at the desired altitude, it is more comfortable for pilot and passenger, tends to keep the engine cooler, and covers a greater distance during climb.

Cruise performance chart—A chart in aircraft flight manuals showing the various cruise speeds which can be maintained at various altitudes with varying rpm and manifold-pressure settings. It also shows fuel consumption and maximum range for each combination.

Cruising altitude—Altitudes at which aircraft should cruise under various conditions, according to the hemispherical rules.

Cruising speed—Any speed in level flight, usually resulting from a power setting recommended in the appropriate cruise-performance chart.

CSDRBL—Considerable.

CST—Coast.

CS/T—Combined station and tower facility.

CTC—Contact.

CTL—Control.

CTLD—Controlled.

Cuban eight—A maneuver in which the airplane completes about three fourths of a normal loop, rolls over and repeats the loop portion in the other direction, resulting in a vertical figure-eight pattern.

Cumuliform—Clouds having rounded or dome-shaped upper surfaces with some protuberances. They usually have flat bases and are formed by rising convection currents in unstable air, and occur separated from each other by downward currents, leading to turbulence in the vicinity of the clouds.

Cumulonimbus—A cumuliform cloud with a dark base and extensive vertical development, usually producing thunderstorms. Also called *thunderheads*.

Cumulus—Cumuliform cloud which develops vertically from a low flat bottom to a billowing top. It is not as high as a cumulonibus.

Customs facilities—Facilities available at international airports, airports of entry and, upon prior notification, at certain other airports, to check international flights into and out of the country. When returning to the U.S. from another country, it is advisable to time the arrival to coincide with the hours during which customs facilities are staffed by on-duty personnel, as overtime charges can be rather steep.

CVR—Cover.

Cyclic pitch control—The control stick of a helicopter used to induce pitch or roll.

Cyclone—A low-pressure area.

Cyclone—A storm occurring in a low-pressure area.

Cylinder—The hollow tube-like structure in a reciprocating engine in which the piston moves back and forth. It forms the circular wall of the combustion chamber.

Cylinder-head temperature—The temperature generated by the combusion chamber.

Cylinder-head-temperature gauge—(CHT)—An engine instrument which indicates the heat being produced by the combustion chamber in reciprocating engines. Usually calibrated in degrees F. A green arc indicates the normal operating temperature range and a red line or arc indicates that the engine is overheating. Overheating of a reciprocating engine is usually the result of pre-ignition or detonation and can produce catastrophic engine failure within a very short period of time.

D—Day (instrument approach charts.)

D—Delta (in the phonetic alphabet.)

D—Dust (in sequence reports.)

DABRK—Daybreak.

DABS—Discrete-address beacon system.

Dakota—A single-engine aircraft manufactured by Piper Aircraft Corporation.

DALGT—Daylight.

Dassault—See Avions Marcel Dassault-Breguet.

Data-link—A phrase used to describe technology in which computers communicate with one another, issuing and accepting commands.

Dauphin—A family of turbine helicopters manufactured by Aerospatiale in France.

dB—DeciBel.

DBA—Doing business as...

DC—Direct current (electric).

DCE—Distance calculating equipment.

DCR—Decrease.

Dead reckoning—A method of navigation by which the course and time of an aircraft between two given points is estimated by taking course, speed and wind components calculated with a wind triangle into consideration. Applicable only in VFR conditions. The phrase "dead" has nothing to do with death, but is a bastardization of the terms *deduced reckoning*.

Dead-stick landing—A landing made without engine power, usually only in an emergency.

Decathlon—One of the Champion family of sport aircraft manufactured by Bellanca Aircraft Corporation.

Piper Dakota (courtesy of Piper).

Deceleration—A decrease in velocity and the rate at which such decrease takes place.

DeciBel—A unit for expressing the relative intensity of sound on a scale from zero for the average least perceptible sound to about 130 for the average sound intensity producing physical pain.

Decision height—(DH)—The altitude at which a pilot making a precision instrument approach (ILS or PAR) must have the airport in sight in order to legally continue the approach. If the airport is not in sight, he must make a missed approach.

Aerospatiale Dauphin (courtesy of Aerospatiale).

109

Decathlon (courtesy of Bellanca).

Decoder—A device used in ATC radar operations to differentiate among signals received from transponders, resulting in their display as selected codes on the radar scope.

Ded Reckoning—Dead reckoning.

Deepening—Increasingly low air pressure in a moving low, usually in its center.

Defense visual flight rules—(DVFR)—Rules applicable to operations within an air-defense identification zone (ADIZ) under visual flight rules.

Defiant—A push-pull twin-engine four-seat aircraft designed and built by Burt Rutan and distinguished by its canard wings.

Deflation port—The opening in a hot-air or gas balloon which can be operated by the pilot and, when opened, permits hot air or gas to escape, thus either stopping ascent or producing descent.

DEG—Degree.

De-icing equipment—See anti-icing equipment.

Delay indefinite because (reason), expect approach/further clearance at (time)—Phrase used by ATC to inform the pilot of a delay for which an accurate time estimate is not available.

Delco Electronics Division of General Motors—Manufacturers of, among other products, long-range navigation systems. (700 East Firmin Street, Kokomo, IN 46901.)

Delta—In aviation-radio phraseology the term for the letter D.

DEMOL—Demolition.

DEMSND—Decommissioned.

Denalt computer—A small computer used to determine density altitude.

Density altitude—Pressure altitude corrected for prevailing temperature conditions. Awareness of density altitude is important in calculating takeoff distance and climb performance, especially when operating to or from an airport at a high elevation. The following table shows the density altitude at various elevations under given temperature conditions.

Density altitude at given temperature conditions and different elevations.

degrees C./F.

Airport elevation	36/97	32/90	28/82	24/75	20/68	16/61	12/54	8/47	4/39	0/32	-4/25
Sea level	2500	2100	1600	1100	600	100	-400	-900	-1400	-1900	-2300
500	3100	2700	2200	1800	1300	800	300	-200	-700	-1200	-1600
1000	3700	3200	2800	2400	1900	1400	900	500	-100	-600	-1100
1500	4300	3900	3400	3000	2500	2000	1500	1100	600	100	-500
2000	4900	4500	4000	3600	3100	2600	2200	1700	1200	700	200
2500	5500	5100	4600	4200	3700	3200	2800	2300	1800	1300	800
3000	6300	5700	5200	4800	4300	3800	3400	2900	2500	2000	1500
3500	6800	6300	5900	5400	4900	4400	4000	3500	3100	2600	2100
4000	7500	7000	6600	5900	5500	5000	4600	4100	3700	3200	2700
4500	8000	7500	7100	6500	6000	5500	5100	4600	4200	3600	3200
5000	8500	8000	7600	7100	6500	6100	5600	5200	4700	4200	3700
5500	9100	8600	8200	7700	7100	6700	6200	5800	5300	4700	4300
6000	9700	9200	8800	8300	7800	7300	6900	6400	5900	5300	4900
6500	10300	9800	9400	8800	8400	7800	7400	7000	6400	6000	5600
7000	10800	10400	10000	9500	9000	8500	8100	7600	7100	6700	6200
7500	11400	11000	10600	10200	9700	9200	8700	8300	7800	7300	6800
8000	12100	11600	11200	10800	10300	9800	9200	8800	8400	7900	7500

Defiant (courtesy of Peter Garrison).

Burt Rutan's Defiant specifications.

BURT RUTAN'S DEFIANT	PROTOTYPE (actual)	PRODUCTION MODEL (estimated)
ENGINES	L/O-320 (2)	L/IO-360-B (2)
TBO (hours)	2,000	2,000
GEAR	fixed	fixed
PROPELLER number of blades, each	2	2
TAKEOFF ground roll (ft)	1,200	850
50' obstacle (ft)	1,500	1,300
RATE OF CLIMB (fpm)	1,600 (1 eng. 300)	1,850 (1 eng. 580)
Vx (knots)	90	85
Vxse (knots)	68	66
Vy (knots)	115	110
Vyse (knots)	78	75
SERVICE CEILING (ft)	22,400 (1 eng. 5,900)	27,000 (1 eng. 10,300)
MAX SPEED sea level (knots)	196	211
CRUISE SPEED AT 70% AT 8,000 ft. (kts)	186	201
AT 55% AT 12,000 ft. (kts)	173	189
RANGE, 45-min res. 70% 8,000 ft (nm)	860	930
55% 12,000 ft (nm)	1,000	1,080
FUEL FLOW per eng. at 70% (pph)	50.9	57.3
at 55% (pph)	41.2	46.3
STALL clean (knots)	65	65
dirty (knots)	65	65
LANDING ground roll (ft)	980	980
50' obstacle (ft)	1,500	1,500
ACCELERATE-STOP DISTANCE (ft)	2,250	2,100
RAMP WEIGHT (lbs)	2,900	3,350
TAKEOFF/LANDING WEIGHT (lbs)	2,900	3,350
USEFUL LOAD (lbs)	1,290	1,550
FUEL CAPACITY (lbs)	540	660
MAX PAYLOAD (lbs)	950	1,100
PAYLOAD w. full fuel (lbs)	750	890
FUEL w. max payload (lbs)	340	450
WING LOADING, gross (lbs per sq ft)	22	22
WING AREA (sq ft)	127	152
LENGTH/HEIGHT/SPAN exterior (ft)	22.9/8.3/29.22	25/8.3/32
LENGTH/HEIGHT/WIDTH cabin (ft)	8/3.6/3.8	8/3.8/4.2

Balloon with deflation port partly open.

Departure control—An ATC service which monitors and directs IFR and in some instances VFR traffic at a controlled airport.

Departure leg—The airborne path of an aircraft immediately after takeoff.

Departure time—The time at which an aircraft becomes airborne.

DEP CON—Departure control.

Depression—A low.

Depth perception—The ability to judge distances with reasonable accuracy.

DER—Designated engineering representative of the FAA.

Descending turn—A turn made during a descent, usually to a predetermined heading, using power appropriate to maintain constant airspeed and rate of descent.

Designated engineering representative—A person designated by the FAA and authorized to issue supplementary type certificates.

Detonation—The burning of a fuel-and-air mixture by explosion rather than steady burning. It tends to occur when the wrong grade of fuel is used and results in rapidly rising cylinder-head temperatures. Prolonged detonation will bring about catastrophic engine failure.

DEV—Deviation.

Deviation—The error in the reading of a magnetic compass induced by installation or by magnetic disturbances in the aircraft. It is the difference between the magnetic heading and the compass reading. Such errors are recorded on the compass-deviation card which must be displayed in the cockpit.

Deviation card—Compass correction card.

DeVore Aviation—Manufacturers of floats for amphibious aircraft. (125 Mineola Avenue, Roslyn Heights, New York 11577.)

DEW—Distant early warning.

DEWIZ—Distant early warning identification zone.

Dew point—The temperature to which air must cool in order for condensation to take place without change in pressure or vapor content.

DF—Direction finder.

DF approach procedure—This procedure is used under emergency conditions when an alternate instrument procedure is unavailable or cannot be executed. Such DF guidance is given by ATC facilities with DF capability.

DF fix—The geographical location of an airplane in flight, obtained by using one or more direction finders.

DF guidance—DF steer.

DF steer—When giving a DF steer to an aircraft, a facility equipped with DF capability gives the aircraft specific headings to fly which, when followed, will guide the aircraft to an airport, back on its course, or to any other predetermined point. DF steers are given aircraft in distress and other aircraft upon request by the pilot. Pilots are encouraged to request practice DF steers as controllers and/or flight service specialists are supposed to give a certain number of DF steers over a fixed period of time in order to stay proficient. Practice DF steers are given on a workload-permitting basis.

DFUS—Diffuse.

DG—Directional gyro compass.

DH—Decision height.

Digital computer—A type of electronic computer which translates all information input into combinations of two digits (zero and one or plus and minus or positive and negative), performs the necessary calculations and then translates its two-digit language into whatever comprehendable type of readout is desired. Today, virtually all computers used in aviation are of the digital type.

Digital timer (courtesy of Poyntek).

Digital timers and clocks comparison chart.

	MANUFACTURER	MODEL	PRICE uninstalled	POWER aircraft elect. syst	POWER separate battery	time of day hrs.	elapsed time to	count up to	count down from	no. of displays	displays month and date	automatically activated by external signal (OM)	visual Alarm	audio	weight lbs.	REMARKS
CLOCKS AND TIMERS	ASTROTECH	LC-2	129.50	OPT		12/24	24 HRS			1					.25	
	DAVTRON	COUNT-DOWN TIMER 822	185.00	14V 28V*				9.59 MINS	9.59 MINS	1			OPT		.4	*REQUIRES RESISTOR
	H.T. INSTRUMENTS	MTI	299.00	14V 28V		24		100 MINS	100 MINS	1					.8	
	POYNTEK	ACCU-NAV	325.00	14V 28V*	9V **	12/24		11:59 23:59	:59	:59					.4	*REQUIRES RESISTOR **BACK-UP
	ASTROTECH	LC-6	349.00	OPT		12/24	24 HRS		24 HRS						.25	IF
	DAVTRON	811B	379.00	14V 28V*		24	24 HRS	24 HRS		2					.6	AIRCRAFT BATTERY IS REMOVED

Digital timers—Timers and clocks with no moving parts which display time or elapsed time in digital form.

Digitizer—The instrument which reads the altitude off the altimeter and transmits that information to the transponder.

Dihedral—The angle at which the wing is attached upward from the horizontal axis of the airframe. Also applies to the horizontal tail and other airfoils.

Direct—Flight in a straight line between two points or fixes. When IFR pilots fly off airways direct, any point used to define the direct-route segments becomes a compulsory reporting point unless the aircraft is in radar contact.

Direct entry—One of the three recommended (but not compulsory) means of entering a holding pattern. In a direct entry the pilot flies along one leg of the holding pattern directly toward the fix and, when passing the fix, turns right and flies the pattern.

Directional gyro—A gyroscopic flight instrument which, when set to conform with the magnetic compass, will continue to indicate the aircraft heading for some time. It tends to gradually develop heading errors (preset) and must be adjusted intermittently by the pilot.

Direction finder—(DF)—A direction finder is a radio receiver which is equipped with a directional sensing antenna capable of taking bearings on a radio transmitter when that transmitter is being activated. When using DF to guide a pilot, the controller or flight service specialist will ask him to depress the transmit button on his mike for a given number of seconds which then enables him to determine the position of the aircraft. DF services are provided by control towers and FSSs listed in Part 2 of AIM.

Dirigible—A lighter-than-air craft equipped with a power source and a means of directional control.

DISCNTD—Discontinued.

Discrete address beacon system—(DABS)—The DABS surveillance system is similar to ATCRBS in that there are ground interrogators and airborne transponders. DABS, however, provides garble-free replies, better quality data, and a means of implementing a digital data link. Each DABS interrogation is addressed to one specific aircraft which then replies only after it has recognized its own address rather than replying to all interrogators within line of sight.

Discrete code—Any one of the 4,096 codes which can be selected in a Mode 3-A transponder except those ending in two zeroes. (0020; 2301; 4356 etc.) Non-discrete codes are used by radar facilities which are incapable of discrete decoding and some are reserved for special use: 1200/VFR; 7600/two-way communication radio failure; 7700/emergency; 3500/being hijacked.

Discrete frequency—A separate frequency being used by a controller in dealing with one or a limited number of aircraft, thus reducing frequency congestion.

Displaced threshold—The touchdown point located on a runway at a point other than the beginning of that runway.

Dissymmetry of lift—The difference in lift produced by the advancing and retreating motion of a helicopter blade during each rotation cycle.

DIST—District.

Distance calculating equipment—A cockpit instrument developed by Collins Avionics Division which is capable of calculating distance to station, time to station and ground speed, as long as the aircraft is within reception distance of two VORs. Unlike DME, it can operate with standard VORs and does not require that the nav aids be VORTACs. It uses input from standard on-board VOR receivers to effect its computation and the digital readout shows distance or time to the station, ground speed, or the bearing to either station. While capable of performing most functions normally obtained from a DME, it cannot be legally used during instrument approaches which require a DME.

Distance from VOR—By flying at a distance past a VOR the distance from the VOR can be calculated by noting the time it takes to pass a given number of radials, assuming the ground speed is known. Or ground speed can similarly be calculated if the distance from the VOR is known. The formula for this computation looks like this:

Time between radials in seconds divided by degrees of radial change equals minutes to the station. Or, ground speed (if not known, use TAS) times the time in minutes between radials divided by the number of degrees of radial change equals miles to the station. (Statute miles if speed is figured in mph, nautical miles if speed is figured in knots.)

Distance measuring equipment—(DME)—A combination of airborne and ground equipment which gives a distance reading from the airplane to the ground station by measuring the time lapse of a signal generated in the airplane and bounced off the ground station. The recorded distance is the slant range in nm from the airplane to the station. The difference between the slant range and the actual horizontal distance is of no

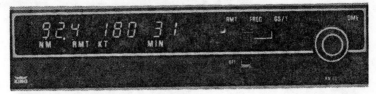

Digital DME (courtesy of King Radio).

practical consequence at altitudes below 18,000 feet when approximately 10 or more miles from the station. When closer to the station, both distance and ground-speed readings become unreliable. Some very sophisticated and expensive DMEs, designed primarily for use in airliners and corporate jets, automatically compute the slant range and produce reliable readouts regardless of altitude and distance from the station.

Ditching—Making an emergency off-airport landing.

Dive—A steep descent with or without power at a speed greater than normal.

Dive brakes—Spoiler-type systems in the wings of some jet aircraft and all gliders which, when extended, produce drag and, in turn, reduce airspeed.

DL—Direct line (interphone).

DLA—Delay.

DME—Distance measuring equipment.

DME fix—A geographical point determined by reference to DME instrument readings. Shown on instrument approach charts in terms of distance in nm and radial degrees from the appropriate nav aid.

DME separation—Separation of aircraft in terms of nm determined by using DME.

DMSH—Diminish.

DN—Day and night (on instrument approach charts.)

DNS—Dense.

DNSLP—Downslope.

DO—Ditto.

DOD—Department of Defense.

DOD FLIP—Department of Defense Flight Information Publications.

Doldrums—The general circulation pattern of a calm or light and variable wind around the equator between the two regions of tradewinds.

Dope—A glue-like varnish used to fill the weave and weatherproof fabric surfaces of aircraft.

Doppler effect—A change in the frequency with which sound, light or radio waves from a given source reach an observer when the source and the observer are in rapid motion with respect to one another, so that the frequency increases or decreases according to the speed with which the distance is increasing or decreasing. (Named after Christian J. Doppler.)

Distance measuring equipment comparison chart.

MANUFACTURER	MODEL	PRICE uninstalled	Volt input	Display range nm	Accuracy nm	Time to station minutes	Ground speed knots	Channels	Display: digital / electric / mechanical / gas discharge	Hold feature	RNAV compatible	Weight lbs.	REMARKS
COLLINS	DCE 400	1,180	14 28	199		120	399				OPT	5.2	TSO. OPERATES WITH INPUT FROM ANY TWO IN-RANGE VORS OR VORTACS. CANNOT BE USED FOR DME INSTRUMENT APPROACHES.
NARCO	DME 190	3,075	14 28	199.9	+/- 0.1	89	400	200				2.4	
KING	KN 62	3,100	14 28	389	+/- 0.1* 1.0**	99	999	200		OPT		9.2	* to 99 nm ** to 389 nm
KING	KN 65A	3,670	14 28	199.9	+/- 0.5	99	400	200				11	TSO.
ARC	RTA 400	3,695	28	200	+/- 0.2	60	400	200				NA	
KING	KN 63	3,850	11 to 30	399	+/- 0.1	99	999	200				8.4	TSO. SIMULTANEOUS READOUT OF GROUND SPEED, DISTANCE AND TIME TO STATION
NARCO	DME 195	3,925	14 28	199.9	+/- 0.1	89	400	200				6.2	
COLLINS	DME-451	3,980	14 28	199.9	+/- 0.1	120	399	200				11	TSO.
ARC	RTA 800	3,995	28	200	+/- 0.2	60	400	252				14.7	
EDO-AIRE/ FAIRFIELD	RT888/ CD 888	3,995	14 28	192	+/- 0.1	999	800	200				8.9	ELAPSED TIMER AND STOP-WATCH FUNCTION INCLUDED
BENDIX AVIONICS	DME 2030	4,616	14 28	200	+/- 0.1	192	450	200				14	TSO.
KING	KDM705A	4,830	28	399.4	+/- 0.1	99	999	200					TSO.
COLLINS	DME-40	6,605	28	250	+/- 0.1	99	999	252				8.1	TSO.
KING	KDM 7000B	8,958.50	110AC 400 Hz	399	+/- 0.1	79	799	252			•	17.1	ARINC CHARACTERISTIC 568 • $80 EXTRA

Doppler radar—A radar system which uses the Doppler effect for measuring velocity.

DOT—Department of Transportation.

Downdraft—A convection current which moves downward. Popularly and falsely called *air pocket*.

Downwind—Moving in the direction in which the wind is blowing; or being located on the lee side of a mountain or other geographic point.

Downwind landing—A landing made with the wind rather than into it and used only in emergencies or occasionally under light wind conditions in order to expedite traffic. It results in a longer landing roll and, under adverse conditions, can cause the aircraft to run off the end of the runway.

Downwind leg—The flight path parallel to the landing runway in the direction opposite to landing. It is part of the standard traffic pattern.

DP—Deep.

DPNG—Deepening.

DPTG—Departing.

DPTR—Departure.

Drag—The force created by the friction of the air on objects in motion. It must be overcome by thrust in order to achieve flight parallel to the relative wind. There are two types of drag, induced drag and parasite drag. Induced drag is drag created through the process of producing lift. Parasite drag is all drag from surfaces which do not contribute to lift. It increases with an increase in airspeed.

Drag chute—A type of parachute which can be deployed from the rear of an aircraft in order to reduce the landing roll.

Dragging—The lead or lag capability of each helicopter rotor blade in relation to each other rotor blade.

DRFT—Drift.

Drift—The gradual displacement of an airplane from its course as the result of a steady crosswind component.

Drift angle—The angle off course corrected by the wind-correction angle.

Drizzle—Precipitation of tiny droplets, usually from stratus clouds. Shown in sequence reports as L.

Droop stops—Means of preventing articulated helicopter rotor blades from drooping too close to the ground during start-up and shut-down.

Drop line—A rope, usually approximately 150 feet long, carried in balloons and dropped to the ground to slow lateral movement or to facilitate assistance in landing by the ground crew.

DRZL—Drizzle.

DSB—Double sideband.

DSIPT—Dissipate.

dsnt—distant.

Dual control—Having a double set of controls in the cockpit, permitting either pilot or co-pilot (or instructor or student) to control the aircraft.

GULFSTREAM-AMERICAN CORP. MODEL:	HUSTLER 500
ENGINES	PW/PT6A-41 (front)
	PW/JT15D-1 (rear)
shp (front)/thrust (lbs) (rear)	850/2,200 (flat rated)
TBO (front)/(rear)	2,500/3,000
SEATS	7
TAKEOFF ground roll (ft)	1,370
50' obstacle (ft)	1,630
RATE OF CLIMB 2 engines (fpm)	3,650
1 engine front/rear (fpm)	1,400/1,450
Vxse front/rear (knots)	125/165
Vyse front/rear (knots)	145/170
Vmca (knots)	75
ACCELERATE-STOP DISTANCE (ft)	3,000
SERVICE CEILING 2 engines (ft)	41,000
1 engine front/rear (ft)	23,000/31,000
MAX SPEED (knots)	400 (25,000 ft)
BEST CRUISE SPEED (knots)	350 (40,000 ft)
LONG-RANGE CRUISE SPEED (knots)	260
RANGE w. res. best speed (nm)	2,000
long-range speed (nm)	2,500
FUEL FLOW best speed, both engines (pph)	510
long range, front engine (pph)	240
STALL clean/dirty (knots)	100/85
LANDING ground roll (ft)	930
50' obstacle (ft)	1,500
RAMP WEIGHT (lbs)	9,580
ZERO FUEL WEIGHT (lbs)	7,176
TAKEOFF/LANDING WEIGHT (lbs)	9,500/9,025
USEFUL LOAD (lbs)	4,572
FUEL CAPACITY (lbs)	3,236
MAX PAYLOAD (lbs)	2,168
PAYLOAD w. full fuel (lbs)	1,336
FUEL w. max payload (lbs)	2,404
WING LOADING (lbs per sq ft)	49.8
WING AREA (sq ft).	190.8
LENGTH/HEIGHT/SPAN external (ft)	41.25/13.2/34.4
LENGTH/HEIGHT/WIDTH cabin (ft)	18.33/4.17/4

Duchess—A light-light twin-engine aircraft manufactured by Beech Aircraft Corporation.

Duke—A pressurized piston-twin aircraft produced by Beech Aircraft Corporation.

Duplex—Simultaneously using two frequencies, one for transmitting, the other for receiving.

DURG—During.

Duster Sailplanes—Manufacturers of sailplanes and sailplane kits. (12676 Pierce Street #8, Pacoima, CA 91331.)

Beechcraft Duchess 76 (courtesy of Beech).

Duty priorities—The order in which services must be provided by ATC controllers and FSS specialists.

ATC controller priorities:

1) Separation of aircraft and issuing radar safety advisories.
2) Other required services not involving separation of aircraft.
3) Additional services to the extent possible.
4) Priorities with reference to emergencies must be left at the controller's discretion because of the great variety of possible emergency situations.

Beechcraft Duke (courtesy of Beech)

FSS specialist priorities:

1) Action involving emergencies in which life or property may be in immediate danger.
2) Action required because of nav aid malfunctions.
3) Service to airborne aircraft.
4) Weather observations and PIREPs.
5) Preflight pilot briefings.
6) Unscheduled broadcasts.
7) Teletype operations.
8) Transcribed weather broadcasts and pilot automatic telephone weather answering service.
9) Scheduled broadcasts.

DVFR—Defense visual flight rules.
DVLP—Develop.
DX—Duplex.

E—East.

E—Eastern Standard Time.

E—Echo (in the phonetic alphabet).

E—Equatorial air mass.

E—Estimated (in sequence reports).

E—Sleet (in sequence reports).

EAA—Experimental Aircraft Association.

EAS—Equivalent airspeed.

Echo—The term used in aviation-radio phraseology for the letter E.

Eddy—A local whirling current of air different from the general flow, usually given in feet msl.

Edo-Aire Fairfield—Manufacturer of a wide variety of avionics equipment. (216 Passaic Avenue, Fairfield, NJ 07006.)

Edo-Aire/Mitchell—Manufacturer of a family of autopilots and flight-control systems. (Mineral Wells, TX 76067.)

Edo-Aire/Seaplane Division—Manufacturer of floats for amphibians and seaplanes. (65 Marcus Drive, Melville, NY 11746.)

EDT—Eastern Daylight Time.

EFCTV—Effective.

EGT—Exhaust gas temperature gauge.

EHF—Extremely high frequency (30,000 to 300,000 MHz).

Elastomeric bearings—Bearing consisting of bonded elastomers and metal in sandwich layers, said to be exceptionally maintenance free.

Electronique Aerospatiale—Manufacturer of a variety of avionics equipment (B.P. No. 4, 93350 Aeroport le Bourget, France.)

ELEV—Elevation.

Elevation—The height of airports, terrain features or manmade objects on the ground, generally given in feet msl.

Elevator—A primary control surface attached to the horizontal stabilizer, which can be moved up or down to control the pitch of the aircraft with reference to its lateral axis.

ELSW—Elsewhere.

ELT—Emergency locator transmitter.

Emergency, declaring an emergency—A pilot declaring an emergency for whatever reason, automatically has the right of way over all other aircraft (except balloons). Once the emergency is over, he is expected to make a detailed report about it to the FAA.

Emergency frequency—Civil: 121.5 MHz; military: 243.0 MHz. Both frequencies are continually guarded by most but not all FAA facilities. Also called *guard frequencies*.

Emergency locator transmitter—A transmitter carried aboard the aircraft which is automatically activated by the impact during a crash or by imersion in water. It can also be activated by the pilot. When activated it transmits a wailing tone on the two emergency frequencies, permitting DF-equipped aircraft to determine the location of the transmitter and, in turn, of the aircraft in trouble. ELTs are manufactured by Chromalloy Electronics; Communications Components Corporation; DME Corporation; Dorne and Margolin; Emergency Beacon Corporation; Garrett Manufacturing Ltd.; Leigh Systems; Martech Corporation; Micro Electronics; Narco Avionics; Pacific Communications Corporation.

Emergency radar flight patterns—Triangular flight patterns flown by a pilot to indicate to an ATC controller that an emergency exists. If both transmitter and receiver are inoperative, the pattern is flown to the left. If the transmitter is inoperative but the receiver is working, the pattern is flown to the right. In each instance each leg of the triangular pattern is flown for two minutes with all turns made at one half standard rate. The pattern is to be flown twice and repeated at 20-minute intervals. Tests flown by pilots to determine the effectiveness of this procedure have proved that it is seldom spotted and that it hardly ever works in a radar environment with a great deal of traffic activity.

EMGCY—Emergency.

Empennage—The tail section of an aircraft, including horizontal and vertical stabilizers, rudder and elevators.

Empty operating weight—The weight of an aircraft including all necessary equipment but excluding fuel.

Empty weight—The weight of the aircraft itself including all undrainable fluids. The figure is always given in the aircraft flight manual. Any optional equipment installed must be added to the empty weight in weight-and-balance calculations.

Encoding altimeter—Altitude encoder.

ENDG—Ending.

ENG—Engine.

Engine analyzer—An exhaust gas temperature gauge with probes and readouts for each cylinder of each engine.

Visual emergency signals (courtesy of FAA).

NEED MEDICAL ASSISTANCE-URGENT	ALL OK-DO NOT WAIT WAVE ONE ARM OVERHEAD	CAN PROCEED SHORTLY WAIT IF PRACTICABLE ONE ARM HORIZONTAL	NEED MECHANICAL HELP OR PARTS-LONG DELAY BOTH ARMS HORIZONTAL

USE THROWING MOTION	OUR RECEIVER IS OPERATING CUP HANDS OVER EARS	DO NOT ATTEMPT TO LAND HERE BOTH ARMS ACROSS FACE	LAND HERE BOTH ARMS FORWARD HORIZONTALLY, SQUATTING AND POINTING IN DIRECTION OF LANDING -REPEAT

NEGATIVE (NO) WHITE CLOTH WAVED HORIZONTALLY	AFFIRMATIVE (YES) WHITE CLOTH WAVED VERTICALLY	PICK US UP-PLANE ABANDONED BOTH ARMS VERTICAL	AFFIRMATIVE (YES) DIP NOSE OF PLANE SEVERAL TIMES

	HOW TO USE THEM IF YOU ARE FORCED DOWN AND ARE ABLE TO ATTRACT THE ATTENTION OF THE PILOT OF A RESCUE AIRPLANE, THE BODY SIGNALS ILLUSTRATED ON THIS PAGE CAN BE USED TO TRANSMIT MESSAGES TO HIM AS HE CIRCLES OVER YOUR LOCATION. STAND IN THE OPEN WHEN YOU MAKE THE SIGNALS. BE SURE THAT THE BACKGROUND AS SEEN FROM THE AIR, IS NOT CONFUSING. GO THROUGH THE MOTIONS SLOWLY AND REPEAT EACH SIGNAL UNTIL YOU ARE ARE POSITIVE THAT THE PILOT UNDERSTANDS YOU.
NEGATIVE (NO) FISHTAIL-PLANE	

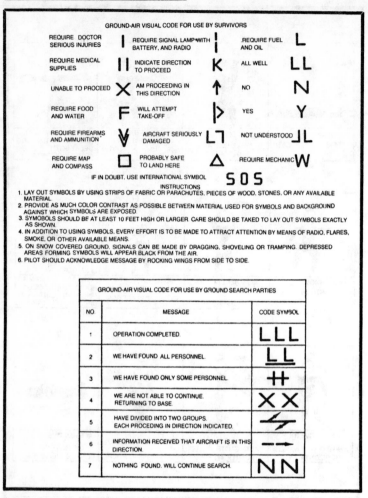

Ground-air visual code for use by survivors and by ground search parties (courtesy of FAA).

Engine instruments—All instruments which indicate condition or performance of the engine, including rpm and manifold pressure. They are usually grouped on the right side of the instrument panel.

Engine, piston—Reciprocating engine, usually with four or more cylinders, burning a mixture of aviation gasoline and air. The average piston engine used in aviation today is air-cooled, horizontally opposed, and generates anywhere from under 100 to 400 horsepower. Most piston engines powering singles and twins are in the 200- to 300-hp range. Normally aspirated (non-turbocharged) piston engines are capable of producing full power only up to about 6,000 feet, above which altitude the power output drops off because of the reduction in the density of the atmosphere. When equipped with turbochargers, full power is available up to two or three times that critical altitude. All pressurized piston-engine aircraft are turbocharged as the turbocharger provides the pressurization capability.

Engine, turbine—All jet engines, pure jets, fan jets and turboprops. For details see jet engines.

Engine, turbocharged—See engine, piston.

En-route air traffic control service—The positive control given aircraft operating on an IFR flight plan, generally by ARTCCs, when such aircraft are in flight between departure and destination airport areas. It functions in the low-altitude as well as the jet-route system.

En-route descent—A gradual descent commenced during the en-route portion of the flight.

En-route flight advisory service—A service which is designed to provide the pilot with weather and other information pertinent to his particular type of flight, route of flight and altitude. It is available at the pilot's request only. Also referred to as *flight-watch service*.

En-route frequency—The frequency used for air-ground communication, usually with center sectors, on or off airways between departure- and approach-control zones.

En-route low-altitude chart—Aeronautical charts showing radio and navigation facilities but no topographical features. Scales vary from 10 to 30 nm per inch. These charts are available from the government or from Jeppesen-Sanderson, Inc. It takes 28 of the government charts to cover the contiguous U.S. and 32 of the Jeppesen charts. The charts can also be obtained from the AOPA or NPA.

Enstrom Helicopter Corporation—Manufacturers of civil helicopters. (Box 277, Menominee, MI 49858.)

Envelope—The hot-air or gas balloon itself, minus gondola and other attachments and systems.

Envelope bag—The bag containing the envelope (balloon) when it is being transported on the ground.

ENTR—Entire.

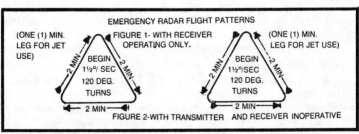

Emergency radar flight patterns (courtesy of FAA).

Enstorm Hawk (courtesy of Enstrom).

EOW—Empty operating weight.

EPNdB—Estimated perceived noise in terms of deciBels.

EQPMT—Equipment.

Equator—The horizontal line around a balloon at a point where the diameter is greatest.

Equatorial air mass—A warm air mass which originates in the doldrums around the equator.

Equilibrium—When flying a balloon, the condition when the lifting capacity of the hot air or gas equals the force of gravity, resulting in constant altitude.

Equivalent airspeed—Calibrated airspeed corrected for compression of air in the pitot system at various altitudes and generally important only at speeds approaching Mach 1. It is equal to the calibrated airspeed at sea level.

ETA—Estimated time of arrival.

ETD—Estimated time of departure.

ETE—Estimated time en-route.

Eustachian tube—A passage between the pharynx and the inner ear, providing humans with a sense of balance. This sense is reliable only when continuously monitored and corrected by visual cues. When no visual cues are available, the sense of balance is disturbed and may cause a pilot, flying in clouds, to believe he is flying straight and level when, in fact, he is in a steep bank.

Evaporation—The process by which water is transformed into vapor, usually by heating the water. The opposite of condensation.

EVE—Evening.

Enstrom Helicopter Corporation piston helicopters specifications.

ENSTROM HELICOPTER COPR. MODEL:	F-28C & F-28C-2	280-C	280-L HAWK
ENGINE	L/HIO-360-E1AD	L/HIO-360-E1AD	L/HIO-360-F1AD
hp	205	205	225
TBO (hours)	1,000	1,000	1,000
SEATS	3	3	4
RATE OF CLIMB, max, sea level (fpm)	1,150	1,150	1,144
SERVICE CEILING (ft)	12,000	12,000	16,000
HIGE (ft)	8,800	8,800	15,500
HOGE (ft)	4,100	4,100	NA
MAX SPEED sea level. (knots)	97	102	102
NORMAL CRUISE SPEED (knots)	93	96	96
Vne (knots)	97	102	102
FUEL CAPACITY (lbs)	240	240	270
RANGE w. full fuel	243 (at 4,000 ft)	243 (at 4,000 ft)	281
GROSS WEIGHT (lbs)	2,350	2,350	2,600
EMPTY WEIGHT (lbs)	1,500	1,500	1,560
USEFUL LOAD (lbs)	850	850	1,040
PAYLOAD w. full fuel (lbs)	440	440	600
w. fuel for 100 nm (lbs)	578	578	774
LENGTH/HEIGHT/WIDTH external (ft)	28.1/9.17/28	29.33/9.22/7.75	32.33/9.45/7.75
cabin (ft)	?/?/5	?/?/4.9	5.8/4/4.9
MAIN ROTOR number of blades	3	3	3
diameter (ft)	32	32	34
TAIL ROTOR number of blades	2	2	2
diameter (ft)	4.67	4.67	4.67
TURBOCHARGER make/type	Rajay 301-E-10-2	Rajay 301-E-10-2	Rajay 301-E-10-2
IFR CERTIFICATION	No	No	No
PRICE (1979 $s)	95.650	99.100	NA

EW—Sleet showers (in sequence reports.)

EWAS—En-route weather advisory service.

EXCP—Except.

Execute missed approach—ATC phraseology used to have the pilot take the following action: Continue inbound to the MAP, then execute the missed-approach procedure as described in the instrument-approach chart. The missed approach procedure and/or any turns associated with it should not be started prior to reaching MAP. In the event that the approach being conducted is a radar approach (ASR or PAR), the missed-approach procedure is to be started immediately after receiving the instruction.

Executing missed approach—Phrase used by the pilot to inform ATC of his intention to execute a missed approach. Alternate phrase: Commencing missed approach.

Exhaust gas temperature gauge and related hardware (courtesy of Alcor).

Executive—A high-performance single-engine aircraft manufactured by Mooney Aircraft Corporation.

Exhaust gas temperature gauge—A means of measuring the exhaust gas temperature from one or several cylinders and displaying the reading in the cockpit. It is accomplished by installing a probe in the exhaust manifold(s) and transmitting the information thus obtained to a gauge located on the instrument panel. Units measuring single cylinders are referred to as EGTs and are always installed in the exhaust manifold of the hottest running cylinder. Units with probes in all cylinder exhaust manifolds are often called *engine analyzers*. EGTs and engine analyzers are the only reliable means of scientifically leaning the mixture of fuel and air to the exact proportion suggested in the engine owners manual, or as desired by the pilot. By operating the aircraft with the leanest mixture permitted by the engine manufacturer at any given altitude, great fuel savings can be realized. When gradually leaning the mixture the EGT needle will climb up to a maximum position. If leaning is continued, it will begin to drop again. The maximum position is referred to as *peak* and represents the mixture at which close to 100 percent of the fuel-air mixture is being burned (but it is not the setting producing best power). Both manufacturers of piston engines (Continental and Lycoming) suggest that the engine be run full rich while operating at 75 or more percent of power. As soon as power is reduced below 75 percent, leaning should commence, regardless of the altitude of the aircraft. Virtually all piston engines can be operated safely at 25 degrees below peak and some may be operated at peak or even at 25 degrees the other side of peak (assuming a power setting below 75 percent). Operating manuals and

Homebuilt aircraft with EXPERIMENTAL placard.

pilot experience with a particular engine should be the guide. EGTs and engine analyzers are manufactured by: Alcor Aviation, Inc.; Avicon; KS Avionics; Radair Division of Terra Corporation; Westberg Manufacturing.

EXPC—Expect.

Expect (altitude) at (time or fix)—ATC phrase used to inform the pilot when to expect clearance to another altitude.

Expect approach clearance at (time)—ATC phrase to inform the pilot, possibly while holding, when to expect clearance for the approach. In instances of an anticipated delay, ATC should advise the pilot at least five minutes prior to reaching the clearance limit of the expected approach-clearance time.

Expect departure clearance at (time)—Used primarily to avoid having aircraft hold on the ground prior to departure and burning fuel while doing so. It refers to the expected time at which the aircraft will be released from the loading gate to commence taxi and takeoff without additional delays.

Expect further clearance at (time)—ATC phrase indicating the time at which the pilot can expect further clearance. If a delay is anticipated, ATC should inform the pilot for the anticipated duration of such a delay at least five minutes prior to his reaching the clearance limit.

Experimental—Any homebuilt aircraft. A placard reading EXPERI-MENTAL must be displayed on homebuilt aircraft and on aircraft which, for one reason or another, are not currently certificated.

Experimental Aircraft Association—(EAA)—The second largest organization of pilots and aircraft owners, comprised primarily of home builders and antiquers.

External loads—Loads carried by a helicopter outside the fuselage. Also referred to as *sling loads*.

EXTRM—Extreme.

EXTSV—Extensive.

Extrusion—A means of forming metal or plastic into any variety of complicated shapes by squeezing it through a form.

F.—Fahrenheit.

F—Fog (in sequence reports.)

F—Foxtrot (in the phonetic alphabet)

FA—Area forecast.

F/A—Fuel-air ratio.

FAA—Federal Aviation Administration.

FAA General Aviation District Offices—Offices concerned with general-aviation matters (GADOs).

Alaskan Region:

    1714 East 5th Avenue, Anchorage, AK 99501

    5640 Airport Way, Fairbanks, AK 99701

    Star Route 1, Box 592, Juneau, AK 99801

Central Region:

    228 Administration Bldg., Municipal Airport, Des Moines, IA 50321

    Fairfax Municipal Airport, Kansas City, KS 66115

    Municipal Airport, Wichita, KS 67209

    9275 Glenaire Drive, Berkeley, MO 63134

    Municipal Airport, Lincoln, NE 68524

Eastern Region:

    National Airport, Washington, D.C. 20001

    Friendship International Airport, Baltimore, MD 21240

    510 Industrial Way, Teterboro, NJ 07608

    Country Airport, Albany, NY 12211

    Republic Airport, Farmingdale, NY 11735

    Monroe Country Airport, Rochester, NY 14517

Allentown-Bethlehem-Easton Airport, Allentown, PA 18103
Harrisburg-York State Airport, New Cumberland, PA 17070
North Philadelphia Airport, Philadelphia, PA 19114
Allegheny County Airport, West Mifflin, PA 15122
Aero Industries, 2nd Floor, Sandston, VA 23150
Kanawha County Airport, Charleston, WV 25311

Great Lakes Region:
P.O. Box H, Dupage County Airport, West Chicago, IL 60185
R.R.2, Box 3, Springfield, IL 62705
St. Joseph County Airport, South Bend, IN 46628
5500 44th Street SE, Grand Rapids, MI 49508
6201 34th Avenue South, Minneapolis, MN 55450
4242 Airport Road, Cincinnati, OH 45226
4393 East 175th Avenue, Columbus, OH 43219
General Mitchell Field, Milwaukee, WI 53207

New England Region:
1001 Westbrook Street, Portland, ME 14102
Municipal Airport, Box 280, Norwood, MA 02062
P.O. Box 544, Westfield, MA 01085

Northwest Region:
3113 Airport Way, Boise, ID 83705
5401 NE Marine Drive, Portland, OR 97218
FAA Building, Boeing Field, Seattle, WA 98108
P.O. Box 247, Parkwater Station, Spokane, WA 99211

Pacific Region:
Room 715, Terminal Building, International Airport, Honolulu,
HI 96819

Rocky Mountain Region:
Jefferson County Airport, Broomsfield, CO 80020
Logan Field, Billings, MT 59101
P.O. Box 1167, Helena, MT 59601
P.O. Box 2128, Fargo, ND 58102
R.R.2, Box 633B, Rapid City, SD 57701
116 North 23rd West Street, Salt Lake City, UT 84116
Air Terminal, Casper, WY 82601
P.O. Box 2166, Cheyenne, WY 82001

Southern Region:
6500 43rd Avenue North, Birmingham, AL 35206
P.O. Box 38665, Jacksonville, FL 38665
P.O. Box 365, Opa Locka, FL 33054
Clearwater International Airport, St. Petersburg, FL 33732

Fulton County Airport, Atlanta, GA 30336
Bowman Field, Louisville, KY 40205
P.O. Box 5855, Jackson, MS 39208
Municipal Airport, Charlottle, NC 28208
P.O. Box 1858, Raleigh, NC 27602
Box 200, Metropolitan Airport, West Columbia, SC 29169
P.O. Box 30050, Memphis, TE 38103
Metropolitan Airport, Nashville, TE 37217

Southwest Region:
Adams Field, Little Rock, AR 72202
Lakefront Airport, New Orleans, LA 70126
Downtown Airport, Shreveport, LA 71107
P.O. Box 9045, Sunport Station, Albuquerque, NM 87119
Wiley Post Airport, Bethany, OH 72008
International Airport, Tulsa, OK 74115
Redbird Airport, Dallas, TX 75232
6795 Convair Road, El Paso, TX 79925
P.O. Box 1689, Meacham Field, Fort Worth, TX 76016
8345 Telephone Road, Houston, TX 77017
P.O. Box 5247, Lubbock, TX 79417
1115 Paul Wilkins Road, San Antonio, TX 78216

Western Region:
2800 Sky Harbor Boulevard, Phoenix, AZ 85034
Air Terminal, Fresno, CA 93727
2815 East Spring Street, Long Beach, CA 90806
P.O. Box 2397, Oakland, CA 94614
International Airport, Ontario, CA 91761
Municipal Airport, Sacramento, CA 95822
3750 John J. Montgomery Drive, San Diego, CA 92123
1887 Airport Boulevard, San Jose, CA 95110
3200 Airport Avenue, Santa Monica, CA 90405
7120 Hayvenhurst Avenue, Van Nuys, CA 91406
5700-C South Haven, Las Vegas, NV 89109
2601 East Plum Lane, Reno, NV 89502
FAA Headquarters: 800 Independence Avenue, Washington, D.C.

FAAP—Federal Aid Airport Program
FAA Standard—Any rule, regulation, standard of comparision, or procedure issued by the FAA, adherence to which is necessary in the interest of the safety of efficiency of the National Airspace System.
FAC—Facility. Also FACIL.
FAD—Fuel advisory departure.
FAF—Final approach fix.

Fahrenheit—(F.)—A temperature scale with 32 degrees as the melting point of ice (freezing point of water) and 212 degrees as the boiling point of water, assuming standard atmospheric conditions at sea level. F. equals 9/5 of the C. reading plus 32 degrees. Currently the predominantly used temperature scale in the U.S., it is gradually being replaced by C. (Celsius, centigrade) as part of the trend toward adopting the metric system.

FAI—Federation Aeronautique Internationale. The international agency supervising record attempts and concerned generally with a variety of standards of international importance.

Fairing—A covering or structure installed on an aircraft to help produce smooth airflow.

Falcon—A family of corporate jet aircraft produced in France by Avions Marcel Dassault-Breguet.

Falcon Jet Corporation—U.S. distributors of the Falcon family of corporate jets. (Pan Am Building, Teterboro Airport, Teterboro, NJ 07608.)

Fan—Popular jargon for propeller.

Fan—Blower (in hot-air ballooning.)

Fanjet—A high-bypass turbine engine in which a fan, for all practical purposes a multi-bladed shrouded propeller, produces a portion of the thrust. Fanjets are quieter than pure jets, but lose a degree of their thrust capability at very high altitudes.

Fan marker—Marker beacon.

FAP—Final approach point.

FAR—Federal Aviation Regulation.

FAS—Flight Assistance Service.

Fasteners—So-called Dzus fasteners used primarily to secure those portions of the cowling which must repeatedly be opened for engine inspection.

Falcon 20.

General Electric CF700 fanjet engine (courtesy of General Electric).

Fast file—A system by which the pilot files a flight plan by phone. It is automatically tape recorded and transmitted to the appropriate ATC facility. Locations having fast-file capability are listed in AIM, Part 3.

Fatique—Pilot tiredness caused by lack of sleep, overwork, excessive noise or other factors.

Fatigue—With reference to metal or aircraft structures the process of wear which may eventually lead to failure.

FAX—Facsimile.

FBO—Fixed base operator; fixed base operation.

FCC—Federal Communications Commission.

FCS—Flight control system.

FCST—Forecast.

FD—Flight director.

FD—Winds aloft forecast.

FE—Flight engineer.

Feather—To change the blade angle of a controllable-pitch propeller so that the blades are turned with the edge facing into the line of flight. It prevents the windmilling of the propeller of an inoperative engine. On helicopters, aligning the rotor blade with the relative wind.

Feathered propeller—See *feather.*

Federal Aviation Administration—(FAA)—The branch of the Department of Transportation (DOT) responsible for the safety, regulation and promotion of civil aviation, and the safe and orderly use of the National Airspace System. Originally an independent agency, called the Federal Aviation Agency, established by the Federal Aviation Act of 1958 during the Eisenhower Administration as a means of upgrading the inefficient Civil Aviation Administration (CAA).

Federal Air Regulations—(FARs)—The basic rules covering the safe and orderly conduct of civil aviation. They are divided into parts:
Parts 1—Definitions and abbreviations.
Parts 11, 13—Procedural rules.

Parts 21 through 59—Aircraft.

Parts 60 through 67—Airmen.

Parts 71 through 77—Airspace.

Parts 91 through 107—Air traffic and general operating rules.

Parts 121 through 137—Air carriers, air taxi operators, air travel clubs, operations for hire, certification and operation.

Parts 141 through 149—Schools and other certificated agencies.

Parts 151 through 169—Airports.

Part 171—Navigational facilities.

Parts 183 through 197—Administrative regulations.

Part 198—War risk insurance.

Part 199—Aircraft loan guarantee program.

Part 430—Accidents and incidents.

Federal Aviation Act of 1958—The federal law which, during the Eisenhower Administration, established the Federal Aviation Agency (FAA).

Federal Communications Commission—(FCC)—The independent agency of the U.S. government charged with the efficient use of radio frequencies. It issues radio-station licenses and operator permits to the aviation community and allots frequencies to the FAA for use in navigation and communication.

Address: Federal Communications Commission,

1919 M Street, NW, Washington, D.C. 20554.

Phone: 202/655-4000.

Feeder airline—A commuter airline serving a limited region and connecting with major airlines.

Feeder route—A flight path shown on instrument approach charts which designates routes to be flown from the en-route structure to the initial approach fix (IAF).

Ferry flight—A flight conducted in order to return an aircraft to its base, delivering an aircraft to another location or moving an aircraft to or from a maintenance facility. Special flight permits can be obtained to ferry a partially disabled but flyable aircraft to another location for repair. No passengers may be carried on such a flight.

FHP—Friction horsepower.

FIDO—Flight Inspection District Office.

Field elevation—Height in feet msl of the highest usable portion of a landing area. When known, may be used to set the altimeter before commencing a flight or to check its accuracy.

FIFO—Flight inspection field office.

Fifteen-meter class—A class of sailplanes with a wingspan of 15 meters or less.

Filed—Used primarily in conjunction with filing an IFR flight plan with ATC. Having filed means operating IFR.

FILG—Filling.

Filling—Changing to a higher air pressure in the center of a moving low.

Fin—A fixed vertical airfoil attached to the fuselage for the purpose of providing stabilization.

Final—Commonly used phrase meaning that an aircraft is on final approach or is aligned with the direction of the active runway.

Final approach—The flight path of an airplane from a specific fix or from its turn after completing the base leg toward the active runway. The third leg of the standard traffic pattern.

Final approach course—A straight line extending from the localizer or runway centerline, without regard to distance from the runway.

Final approach fix—(FAF)—The designated fix from which the final approach may be initiated by an aircraft operating IFR. It identifies the starting point of the final approach segment of an instrument approach.

Final approach point—The point within the published limits of an approach procedure where the aircraft is established on the final approach course and from which the final descent may be started. It is applicable only in non-precision approaches where an FAF has not been established.

Final approach segment—Segments of an instrument approach procedure.

FINFO—Flight Inspection National Field Office.

Fix—A geographical location determined either by visual reference or by electronic nav aids, such as the intersection of two VOR radials or bearings or the DME distance from a nav aid on a given radial or bearing.

Fixed base operation—(FBO)—A commercial operation on an airport which provides such services as fueling, tiedown, maintenance, flight instruction, aircraft sales, restaurant services, etc. or any part thereof.

Fixed base operator—(FBO)—The owner or operator of a fixed base operation.

Fixed-pitch propeller—A propeller, the blade angle of which was set during construction and cannot be changed.

Fixed-wing aircraft—Any heavier-than-air craft other than a helicopter or gyrocopter.

Fixed-wing special IFR—Flight operations, usually conducted by agricultural or industrial aircraft, under a waiver agreement with the FAA. They must be flown by instrument rated pilots in IFR equipped aircraft.

Fixed-wing special VFR—See *special VFR (S/VFR)*.

FL—Flight level.

Flag—A usually red tag incorporated into electronic navigation instruments in the cockpit to alert the pilot to the fact that the instrument is either inoperative or beyond reception distance from the nav aid to which it is tuned.

Flag alarm—See *flag*.

Flameout—Unintentional loss of combustion in a turbine (jet) engine, resulting in loss of power.

Flap—An auxiliary control surface usually located on the trailing edge of the inner wing panels, between the fuselage and the ailerons. It can be extended and/or turned down to increase the wing camber and/or surface, creating additional lift and drag. It is commonly used to increase control during slow flight and to increase the glide angle prior to landing. (See also *Fowler flap; full-span flap*.) Certain high-performance jets and STOL aircraft have leading-edge flaps which can be deployed by the pilot or which deploy automatically at very low speeds.

Flap operating range—The range of airspeeds at which flaps may safely be extended. It is shown on the airspeed indicator by a white arc.

Flapping—The up and down motion of helicopter rotor blades resulting from aerodynamic loads and changes in control input.

Flare—A smooth leveling of the aircraft between glide to a landing and touchdown.

Flare out—Bringing the airplane to its touchdown attitude just inches above the runway.

Flares—Emergency lights which can either be dropped from an aircraft prior to a night emergency landing in order to light up the terrain below, or can be fired from a flare gun or tube by a pilot after an off-airport emergency landing in order to draw attention of passing aircraft or vehicles.

FLASHG—Flashing.

Flashing beacon—Anti-collision light.

FLD—Field.

FLG—Falling.

Flight—Being airborne.

Flight check—In-flight check of navigational aids to determine whether they operate reliably.

Flight computer—A simple device for calculating problems which arise with reference to flight planning, navigation, speeds, fuel consumption etc. It is essentially a circular slide rule.

Flight control system—A combination of flight director, autopilot and other systems, all interlocked to result in near automatic flight.

Flight deck—Cockpit.

Flight director—A computer-equipped flight instrument which collects information from all nav instruments and displays the findings in a pictorial display. It relieves the pilot of the need of scanning all instruments and tells him at a glance what action to take in order to achieve the desired result. When coupled to an autopilot, it becomes a flight control system.

Flight indicator—Gyro horizon.

Flight information region—Airspace of specific dimensions in which flight information and flight alerting services are provided. Information services refer to giving advice and information aimed toward the safe and efficient conduct of flight. Alerting services refer to contacting the CAB

Flight control system (courtesy of Bendix).

Flight control systems comparison chart.

MANUFACTURER	MODEL	PRICE uninstalled	Volt input DC	Volt input AC 400 Hz	Flight director interface	Weight lbs.	REMARKS
EDO-AIRE/MITCHELL	CENTURY IV FD (TQ)	9,886	14 28		CENT. IV FD	25	TWO-CUE OR CROSS-POINTER STEERING HORIZON. LESS HDG. SYSTEM, ARINC. NSD 360A OPT.
EDO-AIRE/MICHELL	CENTURY IV FD (SQ)	10,556	14 28		CENT. IV FD	25	SINGLE CUE OR V-BAR STEERING HORIZON, LESS HDG. SYSTEM, ARINC NSD 360A OPT.
EDO-AIRE/MITCHELL	CENTURY IV FD (4TQ)	11,881	14 28		CENT. IV FD	30	4" TWO-CUE OR CROSS POINTER STEERING HORIZON, LESS HDG. SYSTEM, ARINC NSD 360A OPT.
KING	KFC 200	12,100	14 28		KFC 200	33.9	
BENDIX AVIONICS	FCS 870	15,650	14 28		DH 866A	30	PRICE INCLUDES HSI + FD. YAW DAMPER $2,339 EXTRA
BENDIX AVIONICS	FCS 810	15,900	14 28		DH 841 V/VE	30	
ARC	400B IFCS	15,905	28			31.5	
ARC	800B IFCS	22,025	28			44.5	
KING	KFC 250	23,265	28	115	KFC 250	40.8	INCLUDES ALTITUDE PRESELECT
ARC	1000 IFCS	28,540	28			49.2	
BENDIX AVIONICS	M-4D	35,500	28	115	DH 866A	37	PRICE INCLUDES 4" FD DISPLAY, HSI, REMOTE VG AND YAW
KING	KFC 300	40,251	28	115	KFC 300	69.5	INCLUDES ALT. PRESELECT AND VNAV
SPERRY	SPZ 200	56,580	28	115	SPI-401C 402C,G	29.7	PRICE INCLUDES SPI 401C
SPERRY	SPZ-500/200A	62,885	26 115		SPI-401, 402 501,502	62.2	PRICE INCLUDES SPI-401
SPERRY	SPZ-600	171,570	26 115		SPI-501 502	136.4	PRICE INCLUDES SPI-501 AND ADZ-242 AIR DATA SYST.

The chart additionally includes matrix columns indicating feature availability for each model:

- **Torque limiter:** mechanical, electric
- **Sensors:** direction, rate, vertical
- **Computer capabilities:** air data, a.s. comp., altitude, expand LOC, fault mon, g.s. gain, heading, intercept, navigation, turbulence
- **Other functions:** back crse, glide slp, alt preset, hdg preset, go around, cont. w.st., a.s. hold
- **ADI display**, **HSI display**

a.s. comp. = air speed compensation; expand LOC = expanded localizer; fault mon = fault monitor; g.s. gain = glide slope gain; cont. w.st. = control wheel steering; a.s. hold = air speed hold

Flight director pictorial display (courtesy of Sperry).

and/or other appropriate organizations with reference to an aircraft requiring search and rescue.

Flight inspection—See *flight check.*

Flight instruments—All instruments displaying information about the aircraft's course, speed, altitude and attitude.

Flight level—(FL)—An altitude based on a reference atmospheric pressure of 29.92, stated in digits representing hundreds of feet. It is usually used only for altitudes above 18,000 feet msl. FL 230 means 23,000 feet.

Flight log—A record of a specific flight or of all flight activity by a pilot.

Flight manual—Part 1 of AIM.

Flight path—The actual line, altitude, course along which an aircraft is flying.

Flight path deviation indicator—Course deviation indicator(CDI).

Flight plan—An outline of a proposed cross-country flight, filed with an FSS before departure and closed at the end of the flight. For instrument flight in controlled airspace, flight-plan filing is mandatory. For IFR flights in uncontrolled airspace and for VFR flights, filing a flight plan is voluntary. The only purpose of a VFR flight plan is to activate search and rescue activities in the event that the aircraft fails to arrive at its destination.

Flight plan sequence—The sequence in which flight-plan information should be provided:

1) Type of flight plan (VFR; IFR; DVFR)
2) Aircraft identification number
3) Aircraft type and equipment
4) Estimated TAS
5) Departure point
6) Departure time (Zulu time)
7) Initial cruising altitude
8) Route of flight
9) Destination
10) Remarks
11) Estimated time en-route
12) Fuel on board in hours and minutes
13) Alternate airport (IFR only)
14) Pilot's name
15) Pilot's address or aircraft home base
16) Number of persons on board
17) Color of aircraft

Flight service station—(FSS)—A facility operated by the FAA with the prime responsibility for preflight briefing, communication with VFR flights, assisting lost VFR aircraft, originating and disseminating NOTAMs, broadcasting aviation weather, and operating the national weather teletype system, plus other functions as they arise. (See also *duty priorities*.)

Flight directors comparison chart.

MANUFACTURER	MODEL	PRICE uninstalled	DC	AC 400 Hz	V-bar diam. inch	Autopilot interface	Weight lbs.	REMARKS
KING	KFC 200	12,100				KFC 200	33.7	
COLLINS	FD-112V	18,627	28	115			18.4	COMBINED FLIGHT DIRECTOR AND HSI IN ONE INSTRUMENT
SPERRY	SPI-402G	18,925	28	115	4	SPZ 200	20.5	
COLLINS	FD-108Y	20,256	28	115			20.9	
KING	KFC-250	23,265	28			KFC 250	40.8	
SPERRY	SPI-402C	25,590	28	115	4	SPZ-200	29.7	
SPERRY	SPI-401C	29,120	28	115	4	SPZ-200	31.1	
SPERRY	SPI-402	36,655	28	115	4	SPZ-200A SPZ-500	38.6	VNAV OPT.
KING	KFC 300	40,251	28	115	4	KFC 300	69.5	
SPERRY	SPI-401	40,955	28	26 115	4	SPZ-200A SPZ-500	41.8	VNAV OPT.
COLLINS	FDS-109Z	45,483	28	115	5	AP-105 APS-80	36.5	
SPERRY	SPI-502	45,915	28	26 115	5	SPZ-500 SPZ-600	45.1	VNAV OPT.
SPERRY	SPI-501	47,475	28	26 115	5	SPZ-500 SPZ-600	47.1	VNAV OPT.
COLLINS	FDS-84	N.A.	28	26			N.A.	

cross-ptr = cross pointer; contr. w.st. = control wheel steering; vert speed = vertical speed; ang. of att. = angle of attack; digit. crse = digital course; ground sp. = ground speed; MDA/DH warn = MDA/DH warning; slip indic = slip indicator; time to st. = time to station

143

Flight plan (courtesy of FAA).

Flight service specialist—An employee of the FAA performing his duties in an FSS.

Flight Standards District Office—(FSDO)—A field office of the FAA, serving a given geographical area and primarily serving the aviation industry in that area in matters relating the flight standards, such as the certification of aircraft or systems. Other duties involve certification of airmen, accident prevention, investigation, enforcement of rules and regulations and matters related thereto.

Flight test—The in-flight evaluation of an applicant for a pilot certificate or rating by an authorized examiner.

Flight test—The in-flight investigation of the flight characteristics of an aircraft, an aircraft component or system.

Flight time—The total time from the moment an aircraft first moves under its own power for the purpose of subsequent flight until it comes to rest at the next point of landing. Also known as *block time*.

Flight visibility—The distance it is possible to see while in flight. The visibility from one point in the air to another point in the air.

Flight watch—En-route flight advisory service.

FLIP—Flight information publications.

Flite Liner—A single-engine fixed-gear training aircraft of the Cherokee family, produced by Piper Aircraft Corporation, now superseded by the Tomahawk.

Float—A boat-like structure which replaces the landing gear wheels on a seaplane or amphibian, used to take off and land on water.

Float—The action of an airplane which continues to remain airborne after the flare prior to landing.

Float—A part of the carburetor which measures the amount of fuel required for the correct mixture of fuel and air.

Floater—An aircraft that tends to float after the flare prior to landing.

Floater—A sailplane with an exceptional L/D ratio.

Float plane—Seaplane; usually the type which floats on its hull.

Flow control—Routines designed to regulate the flow of traffic along an airway or other route or en-route to an airport or a given portion of airspace.

FLRY—Flurry.

FLT—Flight.

Flutter—Self-excited oscillations of one or several components of an aircraft, deriving its energy from the airstream. Flutter is dependent upon the elastic and inertia characteristics of the affected component.

FLW—Follow.

FLWG—Following.

Fly (heading/degrees)—An ATC phrase to tell the pilot which heading to fly. If compliance involves a turn, the pilot is expected to turn in the shorter direction to the new heading.

Flying—A monthly aviation magazine with a world-wide circulation of just under a half a million. Published by Ziff-Davis Publishing Company. (See *aviation publications*.)

Flying tail—An empennage structure in which the entire tail assembly is moved as part of the elevator movement.

FM—Frequency modulation.

FM—Fan marker.

FN—Regional forecast.

FOB—Fuel on board.

Foehn gap—In a mountain-wave system marked by clouds, the foehn gap is an area of blue sky between the mountain's cap clouds and the lenticular clouds over the first lee wave.

Foehn wind—A warm, dry wind blowing down the slopes of a mountain range (such as the chinook in the Rocky Mountains).

Interior of a flight service station.

Fog—A cloud formed at or near the surface of the earth, giving a visibility of less than one kilometer (5/8 mile). Fog results from the condensation of tiny water droplets when low altitude air is cooled to or below the dewpoint. Sequence report symbol: F.

Fogged in—Forced to stay on the ground by fog or low visibility caused by any source.

FONE—Telephone.

Forced landing—A landing made necessary by mechanical malfunction, weather, or getting lost at some point other than the intended destination. It may be at an airport or at a location other than an airport. Such forced landings frequently involve short-field technique or the ability to successfully land without the use of power.

Formation flight—Several aircraft, which by prior agreement among the pilots, operate as one with reference to navigation, position reporting, etc. Among aircraft participating in a formation flight separation between aircraft is the responsibility of the participating pilots.

FORNN—Forenoon.

Forward slip—A maneuver in which the aircraft is deliberately cross controlled by using opposite aileron and rudder. It can be used during the final approach in order to achieve rapid loss of altitude, or for the purpose of correcting for wind drift by slipping into the wind.

Foster Airdata, Inc.—Manufacturers of RNAV and VNAV avionics equipment. (7020 Huntley Road, Columbus, OH 43229.)

Four forces—The four primary dynamic forces which make flight possible: Lift, weight, thrust, drag.

Four stroke cycle—The basic operating method of the reciprocating engine; the four movements of each piston: 1) Intake stroke; the movement toward the crankshaft which draws fuel and air into the cylinder. 2) The compression stroke; movement toward the top of the cylinder which compresses the fuel-air mixture for ignition by the spark plug. 3) The power stroke; when pressure of the burning mixture forces the piston back toward the crankshaft which, in turn, is moved by the correcting rod. 4) The exhaust stroke; the piston moves back toward the top of the cylinder which forces what remains of the burned mixture out through the exhaust valve.

Fowler flap—A flap consisting of several individual flaps, capable of initially extending the wing surface without appreciably adding drag, then, when extended further, moving downward, increasing both camber and drag.

Foxjet—A small corporate jet (at this writing in its experimental stage).

Foxjet Industries—Manufacturers of the Foxjet. (6701 West 110th Street, Minneapolis, MN 55438.)

Foxtrot—In aviation-radio phraseology the term used for the letter F.

FP—Flight plan.

fpm—Feet per minute.

FPR—Flight planned route.

Foxjet International jet aircraft specifications.

FOXJET INTERNATIONAL, INC. MODEL:	FOXJET ST/600/8
ENGINES	WR/WR44-800
THRUST (pounds, each)	850
TBO	NA
SEATS	4 to 6
TAKEOFF ground run (ft)	1,600
50′ obstacle (ft)	2,700
RATE OF CLIMB (fpm)	3,400 (1 eng. 1,000)
SERVICE CEILING (ft)	40,000 (1 eng. 25,000)
RANGE (nm)	1,043
CRUISE SPEED (knots)	357
STALL (knots)	74
GROSS WEIGHT (lbs)	4,550
EMPTY WEIGHT (lbs)	2,408
LENGTH/HEIGHT/SPAN (ft)	31.83/10.23/31.62
LENGTH/WIDTH cabin (ft)	12/4.58
PRESSURIZATION (psi)	8.4
PRICE (1979 $s)	795,000
(all performance figures are preliminary)	

FQT—Frequent.

FRACTO—Prefix used with cloud types to indicate that they are broken.

Free balloon—A free-flying balloon, not tethered to the earth.

Freezing rain—Precipitation starting as rain and turning into ice when striking a colder object such as an aircraft or the ground.

FREQ—Frequency.

Frequency—A figure expressing cycles (Hertz) per second of a wave of electromagnetic energy. Radio receivers are designed to select a specific frequency while cutting out all others.

Frequency bands—The radio frequencies utilized in aviation navigation and communication lie in a relatively narrow range of the electromagnetic spectrum between 3,000 cycles (Hertz) and 300,000,000,000 cycles (Hertz). (The total range of the known electro-magnetic spectrum ranges from approximately 85 cycles per second to 100,000,000,000,000,000,000 cycles per second.)

The range applicable to aviations divided into frequency bands, more or less in accordance with their propagation characteristics:

Below 30 kHz	VLF (Very low frequency)
30 kHz to 300 kHz	LF (Low frequency)
300 kHz to 3 MHz	MF (Medium frequency)
3 MHz to 30 MHz	HF (High frequency)
30 MHz to 200 MHz	VHF (Very high frequency)
200 MHz to 3 kMHz	UHF (Ultra high frequency)
3 kMHz to 30 kMHz	SHF (Super high frequency)
30 kMHz to 300 kMHz	EHF (Extremely high frequency)

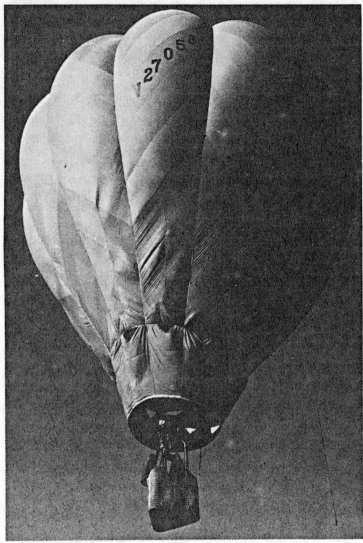

Free hot-air balloon.

The propagation characteristics (reception distance) of these bands breaks
   down as follows:
   VLF........world-wide
   LF, MF.....3,000 to 5,000 miles
   Others.....Line of sight
Frequency drift—The tendency of certain radio receivers to drift from the
   frequency to which they have been tuned.

148

Frequency management system—Sophisticated avionics systems which permit the pilot to tune any number of navigation and communication radios through one central input unit, usually a keyboard.

Frequency Utilization Plan—The specific uses for which certain frequencies are reserved: LF, MF and HF:

190 kHz to 415 kHz and 515 kHz to 544 kHz: L/MF radio ranges, NDBs and ILS compass locators.

535 kHz to 1605 kHz: Standard broadcast stations (AM)

500 kHz: International distress frequency.

VHF:

Navigation

75 MHz: Fan markers, ILS markers, Z markers.

108.0 MHz: Test frequency for VORs.

108.1 to 111.9 MHz (odd tenths only): ILS localizers with or without voice.

108.2 to 111.8 MHz (even tenths only): Terminal VORs.

112.0 to 117.9 MHz: VORs and VORTACs; nav and voice.

Communication

118.0 to 121.4 MHz: ATC

121.5 MHz: Emergency frequency; frequency used by ELTs.

121.6 to 121.9 MHz: Ground control at airports.

122.0 MHz: FSS en-route weather service (flight watch).

122.05 to 122.15 MHz: FSS receive-only frequencies when colocated with a VOR.

122.2 MHz: All FSSs for en-route service.

122.25 to 122.45 MHz: FSS simplex.

122.5 MHz: Tower receive-only frequency.

122.55 to 122.75 MHz: FSS simplex.

122.8 MHz: Unicom at airports without tower or FSS.

122.85 MHz: Available for unicoms at private-use airports.

122.9 MHz: Multicom; air to air; agricultural, parachute operations etc.

122.95 MHz: Unicom at airports with control tower.

123.0 MHz: Unicom at airports with tower or FSS, and as of March, 1979, at uncontrolled airports.

123.05 MHz: Heliport unicom.

123.1 MHz: Search and rescue.

123.15 to 123.55 MHz: Flight tests.

123.3 and 123.5 MHz: Flight schools.

123.6 to 123.65 MHz: Airport advisory by FSS.

123.7 to 128.8 MHz: ATC.

128.85 to 132.0 MHz: Aeronautical radio (ARINC).

132.05 to 135.95 MHz: ATC.

Friction horsepower—The amount of the total horsepower absorbed by engine friction and not available to produce thrust.

Front—The boundary or zone of transition between two air masses with different properties or pressures. Fronts are identified as cold, warm, occluded, stationary.

Frontogenesis—The development of a front, or an increase in intensity of an existing front.

Frontolysis—The dissipation of a front, usually caused by horizontal mixing within the frontal area.

FROPA—Frontal passage.

FROSFC—Frontal surface.

Frost—A layer of crystals of ice formed like dew when the temperature is below 32 degrees F. or when a freezing surface comes in contact with warmer moist air.

FRST—Frost.

FRZ—Freeze.

FRZLVL—Freezing level.

FRZN—Frozen.

FSDO—Flight Standards District Office.

FSL—Full-stop landing.

FSS—Flight Service Station.

FT—Foot, feet.

FT—Fort.

FT—Terminal forecast.

FTHR—Further; farther.

Fuel—A hydrocarbon compound which burns when compressed and mixed with air and then ignited. In doing so it heats and expands, providing the energy necessary to drive the engine.

Fuel advisory departure—Procedures devised to reduce fuel consumption when delays are expected at an airport.

Fuel-air mixture—The mixture of fuel and air channelled into the combustion chamber. A burnable mixture can be classified as lean (excessive percentage of air) or rich (excessive percentage of fuel). The perfect mixture for combustion in the case of aviation gasoline is 15 parts of air to one part of fuel.

Fuel-air ratio—The ratio of fuel to air drawn into the combustion chamber and controlled by the mixture-control knob on the instrument panel.

Fuel consumption—The amount of fuel used by an aircraft per unit of time, usually expressed in gallons or pounds per hour.

Fuel dumping—The release of usable fuel while airborne, usually for the purpose of reducing aircraft weight prior to landing.

Fuel flow—The rate at which fuel is burned by the engine. Usually expressed in gallons or pounds per hour.

Fuel gauge—The cockpit instrument which displays the amount of fuel left in the tank(s). Fuel gauges are historically inaccurate and should never be solely relied on during the last quarter of fuel quantity shown.

Fuel injection—A process, doing away with the carburetor, by which fuel is forced under pressure directly into the combustion chamber.

Fuel monitoring systems—Systems containing a computer and a computer memory. Once told the amount of fuel on board, they keep careful track of the amount of fuel being burned, the amount remaining etc. One of the most popular and reliable fuel monitoring systems is produced by Silver Instruments (1896 National Avenue, Hayward, CA 94545.)

Fuel quantity indicator—Fuel gauge.

Fuel rate—Gallons or pounds per hour.

Fuel siphoning—Unintentional loss of fuel because of a loose fuel cap, leaking tank or such.

Fuel strainer—A filter which removes impurities from the fuel before it reaches the carburetor or injectors. It should be checked during preflight and cleared after refueling.

Fuel venting—Fuel dumping.

Full flaps—The position of the flaps when they are extended as far as possible, providing a maximum amount of additional lift and drag. On some light aircraft it can be dangerous to use full flaps during a crosswind landing.

Full lean—The leanest mixture at which a reciprocating engine will barely continue to operate. Not recommended for prolonged operation.

Full rich—The setting of the mixture control which results in the highest possible proportion of fuel to air which can be supplied to an engine. To be used during takeoff and when operating at 75 or more percent of power.

Full-span flaps—Flaps which either extend the entire length of the wing (such as on the Mitsubishi MU-2 which uses spoilers instead of ailerons) or which continue underneath the fuselage. (See photo under *experimental*.)

Full throttle—Maximum available power at any given altitude.

Fuselage—The main body of the airplane, housing crew, passengers and baggage/cargo space.

FWD—Forward.

FXD—Fixed.

FYI—For your information.

G—Golf (phonetic alphabet)

G—Gusts (in sequence reports)

G—Gravity; the force equal to the gravitational pull of the earth. For instance, in a steep bank an aircraft may produce a three-G pull on the pilot and on the structure of the aircraft itself; a force equal to three times the pull of gravity of the earth.

GI, **G**II, **G**III, Gulfstream I, II and III.

G forces—The gravity forces caused by unusual maneuvers or by rapid acceleration have the following effects on the human body:

3.5 to 4 g loss of peripheral vision; possibly dimming of all vision; greyout.

4 to 4.5 g complete loss of vision; blackout.

Above 4.5 g loss of consciousness.

In order to operate under conditions of greater G forces, the pilot must be wearing a pressure suit.

G forces created by banking the aircraft:

Angle of bank	G load	Increase in stall speed
0	1	0
20	1.065	3
40	1.31	14.4
60	2	41.4
80	5.76	140
90	infinite	infinite

G suit—Pressure suit.

GA—Glide angle.

GADO—General Aviation District Office (FAA).

Gage—Gauge.

GAL—Gallon.

Galley—On-board equipment for food and beverage storage and service.

Learjet 25D (courtesy of Gates Learjet).

GAMA—General Aviation Manufacturers Association.

GAS—Gasoline.

Gas—The lifting agent used in gas balloons; helium, hydrogen or kitchen gas.

Gas balloon—A balloon deriving its buoyancy from gas rather than hot air.

Gascolator—Fuel strainer.

Gasoline—Fuel used in reciprocating engines. Jet engines, though able to function with gasoline in an emergency, normally burn kerosene.

Gate—The passenger loading and unloading position at an air terminal.

Gate hold procedures—Procedures at certain busy airports designed to hold aircraft at the gate or other designated positions when anticipated departure delays exceed five minutes.

Gates Learjet Corporation—Manufacturers of the Learjet family of corporate jet aircraft. (P.O. Box 7707, Wichita, KS 67277.)

Learjet 50-series (courtesy of Gates Learjet).

153

Gates Learjet jet aircraft specifications.

GATES LEARJET CORP. MODEL:	24F	250	28/29 (preliminary)	35A	36A	54/55 (preliminary)
ENGINES	GE-CJ610-8A	GE-CJ610-8A	GE-CJ610-8A	GA/TFE 731-2-2B	GA/TFE 731-2-2B	GA/TFE 731-3
THRUST (pounds, each)	2.950	2.950	2.950	3.500	3.500	3.650
TBO (hours)	4.000	4.000	4.000	Progressive	Progressive	Progressive
SEATS (crew + passengers)	2 + 6 to 9	2 + 8 to 10	2 + 8/10	2 + 8 to 10	2 + 6 to 8	2 + 8 to 11
BALANCED FIELD LENGTH, ISA (ft)	3.297	3.937	2.630/2.880	4.224	4.784	3.994/4.240
ISA + 20 (ft)	3.600	4.775	NA	4.900	5.120	NA
RATE OF CLIMB (fpm)	7.100 (1 eng. 2.050)	5.300 (1 eng. 1.725)	6.195/6.030 (1 eng. NA)	4.900 (1 eng. 1.500)	4.525 (1 eng. 1.325)	NA
Vr (knots)	125	130	NA	137	137	NA
V2 (knots)	126	129	NA	138	138	NA
SERVICE CEILING (ft)	51.000 (1 eng. 27.000)	51.000 (1 eng. 23.500)	51.000 (1 eng. NA)	45.000 (1 eng. 25.300)	45.000 (1 eng. 23.500)	51.000 (1 eng. NA)
Mmo	82	82	82	83	83	82
ALTITUDE CHANGEOVER (FL)	240	240	240	240	240	NA
Vmo (knots)	360	359	359	359	359	NA
CRUISE hi speed (knots)	451 (at FL 430)	447 (at FL 430)	431 (at FL 490)	445 (at FL 430)	435 (at FL 450)	430 (at FL 450)
long range (knots)	421 (at FL 470)	415 (at FL 430)	401 (at FL 490)	414 (at FL 430)	417 (at FL 450)	401 (at FL 450)
RANGE (4 passengers) hi speed (nm)	1.153	1.348	1.150-1.331	2.094	2.520	1.908/2.277
long range (nm)	1.357	1.439	NA	2.310	2.740	2.030/2.411
FUEL FLOW per eng., hi speed (pph)	1.435	1.438	NA	1.085	946	1.016
long range (pph)	1.060	1.214	NA	896	847	873
RAMP WEIGHT (lbs)	13.800	15.500	15.500	17.250	18.250	18.750/19.250
ZERO FUEL WEIGHT (lbs)	11.400	11.400	11.400	13.500	13.500	13.500
TAKEOFF/LANDING WEIGHT (lbs)	13.500/11.880	15.000/13.300	15.000/13.300	17.000/14.300	18.000/14.300	18.500(54) 19.000 (55)/16.000
USEFUL LOAD (lbs)	5.770	6.760	6.205/6.243	7.190	8.146	7.611/8.092
FUEL CAPACITY (lbs)	5.628	6.098	1.684.5.373	6.238	7.440	5.813/6.809
PAYLOAD w. full fuel (lbs)	442	1.162	2.001.1.370	1.202	956	2.048/1.533
WING AREA (sq ft)	58.2	64.7	NA	67.11	71.06	NA
WING LOADING (lbs per sq ft)	231.77	231.77	264.5	253.3	253.3	264.5
LENGTH/HEIGHT/SPAN external (ft)	43.25/12.25/35.55	47.55/12.25/35.55	47.6.12.3.43.8	48.72/12.25/39.5	48.72/12.25/39.5	55.11/14.7/43.79
LENGTH/HEIGHT/WIDTH cabin (ft)	9/4.33/4.95	12.08/4.33/4.95	14.3(28)12.7(29).4.3.4 9	12.95/4.33/4.95	10.75/4.33/4.95	18.3(54) 16.67 (55)/5.7/5.9
PRESSURIZATION (psi)	9.4	9.4	S	9.4	9.4	9.4
EXTERIOR NOISE (EPNdB)	85.8/103.7/95.3	90.9/103.7/95.2		83.6/87.4/91.3	83.6/87.4/91.3	
PRICE (1979 $s)	1.585.000	1.655.000	1.805.000-1.855.000	2.190.000	2.295.000	3.100.000/3.125.000
STANDARD EQUIPMENT:						
VHF nav and com w. AUDIO PANEL	Yes	Yes	Yes	Yes	Yes	Yes
MARKER	Yes	Yes	Yes	Yes	Yes	Yes
GLIDE SLOPE	Yes	Yes	Yes	Yes	Yes	Yes
TRANSPONDER	Yes	Yes	Yes	Yes	Yes	Yes
ENCODING ALTIMETER	Yes	Yes	Yes	Yes	Yes	Yes
ADF, DME, RMI	No	No	No	No	No	No
RNAV/VNAV	Yes	Yes	Yes	Yes	Yes	Yes
AUTOPILOT	Yes	Yes	Yes	Yes	Yes	Yes
FLIGHT DIRECTOR, HSI	Yes	Yes	No	No	No	No
RADAR ALTIMETER	No	No	Yes	Yes	Yes	Yes
WEATHER RADAR	No	No	Yes	Yes	Yes	No
GALLEY	Yes	Yes	Yes	Yes	Yes	Yes
TOILET	Yes	Yes	Yes	Yes	Yes	Yes
De-ICING airfoils	Yes	Yes	Yes	Yes	Yes	Yes
THRUST REVERSER	No	No	No	No	No	No

Gauge—An instrument displaying fuel quantities or other measurements. Also spelled *gage*.

Gazelle—A single-engine turbine helicopter with a shrouded tail rotor, manufactured by Aerospatiale in France.

GCA—Ground controlled approach.

Gear-up landing—A landing during which the gear is either accidentally or purposely kept in the retracted position. Also *belly landing*.

Genave—General Aviation Electronics.

General aviation—All aviation other than military and the airlines.

General Aviation District Office—(GADO)—FAA offices concerned with serving general aviation. For addresses, see FAA.

General Aviation Electronics—Manufacturers of a variety of avionics equipment. (4141 Kingman Drive, Indianapolis, IN 46226.)

General Electric—Manufacturer of turbine engines. (1000 Western Avenue, Lynn, Mass. 01910.)

General Services Administration—The agency of the U.S. government concerned with general housekeeping chores.

General warning signal—An alternating red and green light signal which warns the pilot of a no-radio aircraft to exercise extreme caution.

Generator—A device, identical in construction to an electric motor, which generates electricity and continuously recharges the battery.

GF—Ground fog (in sequence reports.)

GICG—Glaze icing.

Giovanni Agusta S.p.A.—Manufacturers of single-engine turbine helicopters. (C.P. 193 Cascina Costa, Gallarte, Italy. U.S. representative: Atlantic Aviation, P.O. Box 1709, Wilmington, DE 19899.)

Glasflugel—Manufacturers of high-performance sailplanes. (7311 Schlettstall Krs., Nurtingen, West Germany. U.S. representative: Graham Thompson Ltd., 3200 Airport Avenue, Santa Monica, CA 90405.)

Glaze—Clear ice.

Aerospatiale Gazelle (courtesy of Aerospatiale).

155

GLDR—Glider.

Glide—A sustained forward flight in which speed is maintained by sacrificing altitude, power-on or power-off. Every airplane has a glide airspeed depending on load and altitude, which, in the event of engine failure, will enable it to achieve the maximum distance over the ground. For details, see the flight manual of the particular aircraft.

Glide path—The downward flight path of a gliding aircraft.

Glider—Sailplane. An aircraft without an engine. It is supported in flight by the dynamic action of the air on the lifting surfaces. By taking advantage of rising air currents a high-performance sailplane can stay aloft for many hours and cover distances of a thousand miles or more. The world altitude record for a sailplane is over 46,000 feet.

Glide ratio—Usually expressed in terms of L/D (lift over drag) it is the ratio of forward distance traveled to altitude lost assuming still air. An L/D number of 45 (highest achieved by a sailplane to date) means that it will travel 45 miles while losing 1 mile in altitude, assuming still air.

Glide slope—A single-direction VHF radio beam, part of an ILS system, which establishes a glide path at a vertical angle of about 2.5 to 3 degrees relative to the runway. It activates a cockpit display which, when kept centered, shows the pilot that he has achieved the correct rate of descent.

Glide slope—Any proper glide path toward a landing, such as may be shown by a visual approach slope indicator (VASI).

Glide slope intercept altitude—The minimum altitude at which the glide slope must be intercepted on a precision approach in order to assure safe obstacle clearance. It is shown on instrument approach charts.

Gliding distance—The maximum distance a heavier-than-air craft can glide, without power, from any given altitude.

Glitch—Jargon for any trouble or problem, usually of a mechanical nature.

High-performance glider.

Global Navigation—Manufacturers of long-range navigation systems. (24701 Crenshaw Boulevard, Torrance, CA 90505.)

GMT—Greenwich mean time.

GND—Ground.

GNDFG—Ground fog.

Go ahead—Proceed with your message. In aviation-radio practice it may not be used for any other purpose.

Go around—ATC instruction to the pilot to abandon the approach and landing, usually followed by additional instructions.

Going by the book—An expression used by the air traffic controllers to justify slowdowns in intermittent labor disputes with the FAA. Controllers are by law prohibited from calling a strike, therefore they use the slow-down technique instead. The "book" being referred to is the controller's handbook which specifies spacing between aircraft during IFR conditions. When these spacings are extended into VFR conditions, massive delays tend to result at major airports.

Golden Eagle—A pressurized piston twin-engine aircraft manufactured by Cessna Aircraft Company.

Golf—In aviation-radio phraseology the term used for the letter G.

Gondola—The basket of a hot-air or gas balloon.

Goodrich, B.F.—Manufacturer of aircraft tires and brake systems. (500 Main Street, Akron, OH 44318.)

Goodyear Aerospace—Manufacturer of aircraft tires and brake systems. (1210 Massilon Road, Akron, OH 44315.)

Gould Inc.—Formerly Hoffman Electronics Corporation; manufacturers of a variety of navigation equipment. (4323 Arden Drive, El Monte, CA 91731.)

Goulton Industries—Manufacturers of aircraft batteries. (212 Durham Avenue, Metuchen, NJ 08840.)

Cessna 421 Golden Eagle (courtesy Cessna).

Gondola for a hot-air balloon.

GP—Glide path.

gph—Gallons per hour.

GPWS—Ground proximity warning system.

Gradient—Change in value per distance unit, especially of a meteorological element such as pressure gradient.

Gravity—The force exerted by the mass of the earth on any object at or near its surface, pulling the object toward the center of the earth. Roughly equivalent to weight.

Gravity feed—A type of fuel system which depends on gravity to move the fuel from the tanks to the carburetor or injectors.

GRDL—Gradual.

Greasing it on—Making a perfect landing.

Grease job—Perfect landing.

Great circle—A circle on the surface of the earth, an arc of which connecting two terrestial points constitutes the shortest distance between those two points.

Great circle route—A route flown along the great circle or a portion thereof. Flying the great circle route from New York to Paris, for instance, saves 145 miles over the straight (rhumb) line.

Great Lakes—An aerobatic biplane, no longer in production.

Greenwich mean time—Zulu time. The time at the Prime Meridian in England, used in aviation the world over in order to avoid confusion with reference to time zones.

158

Grayout—Blurred vision caused by excessive G forces.

Grimes Manufacturing Company—Manufacturers of aircraft lighting systems. (515 North Russell Street, Urbana, OH 43078.)

Gross weight—The full weight of an aircraft at takeoff consisting of empty weight and useful load. It must not exceed the maximum allowable gross weight for the particular aircraft.

Ground clutter—Patterns produced by ground returns on radar scopes. Such returns tend to degrade other radar returns. Ground clutter can be minimized by using a moving-target indicator (MTI), the effect of which is that the radar will pick up only targets which are in motion.

Ground control—An ATC service at controlled airports, responsible for the safe and efficient movement of aircraft and airport vehicles on the ground.

Ground controlled approach—(GCA)—A non-precision approach using surveillance radar. It makes instrument landings possible for pilots flying aircraft equipped only with a com transceiver and attitude flight instruments by transmitted precise heading and rate-of-descent instructions by voice. When precision approach radar (PAR) is used in a GCA, it becomes a precision instrument approach.

Ground crew—In ballooning, the crew which follows the balloon on the ground and eventually assists in the landing procedure.

Ground effect—A certain amount of additional lift which takes effect when the aircraft is close to the ground. Low-wing aircraft are more susceptible to ground effect than are high-wing aircraft.

Ground handling—The ease or lack of ease with which an aircraft can be maneuvered on the ground.

Ground handling ropes—The ropes attached to a hot-air or gas balloon, used to control its movements during inflation and prior to liftoff.

Ground loop—An uncontrollable abrupt turn of an aircraft while taxiing or during the ground roll of a takeoff or landing. Tailwheel aircraft are specially prone to ground loop during crosswind conditions.

Ground crew unloading gondola.

159

Ground crew assisting in landing.

Ground proximity warning systems—Sophisticated avionics systems capable of warning the pilot when he gets too close to the ground. Usually employing radar technology.

Ground roll—Landing roll.

Ground run—The distance an airplane taking off or landing travels down the runway. The ground run is affected by density altitude, wind, runway-surface conditions and other variables.

Ground speed—(GS)—The speed with which an aircraft travels across the ground. Its relation to TAS is affected by the prevailing winds.

Ground visibility—The horizontal visibility near the earth's surface as reported by an authorized observer. Always given in statute miles, or, in the case of runway visual range (RVR), in feet.

Ground wire—A wire attached to the aircraft from a grounding point during refueling to avoid the possibility of static electricity buildup.

Grumman-American Aviation Corporation—See Gulfstream-American Corporation.

GRVL—Gravel.

GS—Glide slope.

GS—Ground speed.

GSA—General Services Administration.

GTCL—Great circle.

Guard—To listen to. An ATC instruction to guard a given frequency.

Guard frequency—Emergency frequency.

Gulfstream II.

Gulfstream I, II, III—A family of corporate aircraft; the I is a turboprop and no longer in production. The II and III are corporate jets with intercontinental range. Manufactured by Gulfstream-American Corporation.

Gulfstream American Corporation—Manufacturers of a variety of personal, corporate and agricultural aircraft. (P. O. Box 2206, Savannah, GA 31402.)

Gust—An abrupt but brief increase in the velocity of the wind, generally caused by friction between the air and ground features or by uneven heating of the earth's surface.

Gust loads—The load factor caused by gusts and acting upon the aircraft structure.

Gust lock—Devices to hold control surfaces in place while the aircraft is tied down, designed to avoid structural damage caused by gusts of wind.

Gulfstream-American Tiger (courtesy of Gulfstream-American).

Gulfstream-American Corporation jet aircraft specifications.

GULFSTREAM-AMERICAN CORP. MODEL:	GULFSTREAM II w. tip tanks	GULFSTREAM III
ENGINES	RR/Spey Mk 511-8 (2)	RR/Spey Mk 511-8 (2)
THRUST (pounds) each	11,400	11,400
TBO (hours)	7,000	7,000
SEATS crew + passengers	2 + 19	2 + 19
BALANCED FIELD LENGTH, ISA (ft)	5,700	5,550
ISA + 20 (ft)	7,200	7,150
RATE OF CLIMB (fpm)	4,000 (1 eng. 1,200)	4,400 (1 eng. NA)
Vr (knots)	140	139
V2 (knots)	152	142
SERVICE CEILING (ft)	43,000 (1 eng. 24,300)	45,000 (1 eng. 25,000)
Mmo	.85	.85
ALTITUDE CHANGEOVER (FL)	280	280
Vmo (knots)	338	340
CRUISE hi speed (knots)	501 (FL 290)	501 (FL 300)
long range (knots)	430 (FL 410)	445 (FL 430)
RANGE hi speed, max payload (nm)	2,175	2,200
max fuel (nm)	2,550	2,525
long range, max payload (nm)	2,680	3,080
max fuel (nm)	3,180	3,600
FUEL FLOW per engine, hi speed (pph)	3,234.5	NA
long range (pph)	1,491.5	1,285
RAMP WEIGHT (lbs)	66,000	66,000
TAKEOFF/LANDING WEIGHT (lbs)	65,500/58,500	65,500/58,500
ZERO FUEL WEIGHT (lbs)	42,000	42,000
USEFUL LOAD (lbs)	28.814	28,280
FUEL CAPACITY (lbs)	26,800	26,680
MAX PAYLOAD (lbs)	4,814	4,280
FUEL w. max payload (lbs)	24,000	24,000
PAYLOAD w. full fuel (lbs)	2,014	1,600
WING LOADING (lbs per sq ft)	80.9	70.08
WING AREA (sq ft)	809.6	934.6
LENGTH/HEIGHT/SPAN external (ft)	79.92/24.5/72.5	82.92/23.67/77.83
LENGTH/HEIGHT/WIDTH cabin (ft)	39.33/6.08/7.33	41.33/6.08/7.33
PRESSURIZATION (psi)	9.45	9.45
EXTERIOR NOISE (EPNdB)	91/98.2	90/97
PRICE (1979 $s)	NA	NA
STANDARD EQUIPMENT:		
VHF NAV and COM w. OBI	No	Yes
AUDIO PANEL	No	Yes
MARKER	No	Yes
GLIDE SLOPE	No	Yes
TRANSPONDER	No	Yes
ADF	No	Yes
DME	No	Yes
AUTOPILOT	Yes	Yes
FLIGHT DIRECTOR, HSI	Yes	Yes
RADAR ALTIMETER	No	Yes
HF COM	No	Yes
DEICING airfoils	Yes	Yes
THRUST REVERSERS	Yes	Yes
WEATHER RADAR	No	Yes
COCKPIT VOICE RECORDER	Yes	Yes

GWT—Gross weight.

Gyro—Gyroscope.

Gyro compass—See *directional gyro*.

Gyrocopter—A rotary-wing aircraft in which the engine drives a propeller to produce forward speed, and this forward speed causes the rotor to turn, producing lift.

Gulfstream-American Corporation piston aircraft specifications.

GULFSTREAM-AMERICAN CORP. MODEL:	GULFSTREAM II w. tip tanks	GULFSTREAM III
ENGINES	RR/Spey Mk 511-8 (2)	RR/Spey Mk 511-8 (2)
THRUST (pounds) each	11,400	11,400
TBO (hours)	7,000	7,000
SEATS crew + passengers	2 + 19	2 + 19
BALANCED FIELD LENGTH, ISA (ft)	5,700	5,550
ISA + 20 (ft)	7,200	7,150
RATE OF CLIMB (fpm)	4,000 (1 eng. 1,200)	4,400 (1 eng. NA)
V_r (knots)	140	139
V_2 (knots)	152	142
SERVICE CEILING (ft)	43,000 (1 eng. 24,300)	45,000 (1 eng. 25,000)
Mmo	.85	.85
ALTITUDE CHANGEOVER (FL)	280	280
Vmo (knots)	338	340
CRUISE hi speed (knots)	501 (FL 290)	501 (FL 300)
long range (knots)	430 (FL 410)	445 (FL 430)
RANGE hi speed, max payload (nm)	2,175	2,200
max fuel (nm)	2,550	2,525
long range, max payload (nm)	2,680	3,080
max fuel (nm)	3,180	3,600
FUEL FLOW per engine, hi speed (pph)	3,234.5	NA
long range (pph)	1,491.5	1,285
RAMP WEIGHT (lbs)	66,000	66,000
TAKEOFF/LANDING WEIGHT (lbs)	65,500/58,500	65,500/58,500
ZERO FUEL WEIGHT (lbs)	42,000	42,000
USEFUL LOAD (lbs)	28.814	28,280
FUEL CAPACITY (lbs)	26.800	26,680
MAX PAYLOAD (lbs)	4,814	4,280
FUEL w. max payload (lbs)	24,000	24,000
PAYLOAD w. full fuel (lbs)	2,014	1,600
WING LOADING (lbs per sq ft)	80.9	70.08
WING AREA (sq ft)	809.6	934.6
LENGTH/HEIGHT/SPAN external (ft)	79.92/24.5/72.5	82.92/23.67/77.83
LENGTH/HEIGHT/WIDTH cabin (ft)	39.33/6.08/7.33	41.33/6.08/7.33
PRESSURIZATION (psi)	9.45	9.45
EXTERIOR NOISE (EPNdB)	91/98.2	90/97
PRICE (1979 $s)	NA	NA
STANDARD EQUIPMENT:		
VHF NAV and COM w. OBI	No	Yes
AUDIO PANEL	No	Yes
MARKER	No	Yes
GLIDE SLOPE	No	Yes
TRANSPONDER	No	Yes
ADF	No	Yes
DME	No	Yes
AUTOPILOT	Yes	Yes
FLIGHT DIRECTOR, HSI	Yes	Yes
RADAR ALTIMETER	No	Yes
HF COM	No	Yes
DEICING airfoils	Yes	Yes
THRUST REVERSERS	Yes	Yes
WEATHER RADAR	No	Yes
COCKPIT VOICE RECORDER	Yes	Yes

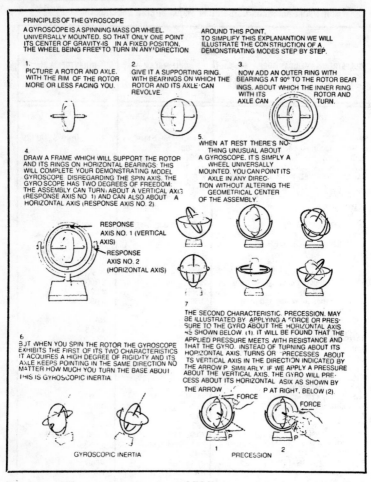

Principles of the gyroscope (courtesy of FAA).

Gyro horizon—See *artificial horizon.*

Gyroscope—A rotating wheel which tends to maintain its position in space relative to a reference direction regardless of the forces exerted on it. It is the basic principle on which the artificial horizon, direction gyro and inertial guidance systems work.

Gyrosyn compass—A compass system combining a remote mounted magnetic compass (usually located near the wing tip or other magnetic-interference-free location) a remote compass transmitter, a directional gyro, an amplifier and the heading indicator. Also referred to as a slaved gyro.

H—Haze (in sequence reports).

H—Hotel (phonetic alphabet).

HAA—Height above airport.

HAA—Helicopter Association of America.

HADIZ—Hawaiian Air Defense Identification Zone.

Hail—Precipitation of irregular balls of ice, often of such size as to be capable of causing considerable damage to aircraft in flight and on the ground, built up by continuously rising and falling in the atmosphere, usually in thunderstorms, until heavy enough to fall to the ground. The sequence report symbol for hail is A.

HAL—Height above landing.

Hammerhead stall—An aerobatic maneuver in which the aircraft climbs almost vertically to the point of stalling and then is tipped right or left in a wingover, ending in a 180-degree change in direction.

Handoff—The action of passing control of, and communication with, an aircraft from one controller to another within the same facility or between facilities without interrupting radar contact.

Hand signals—Signals used by line personnel in directing aircraft on the ground.

Hangar—An enclosed structure for the purpose of parking aircraft or performing maintenance and repair work.

Hangar flying—An expression used for rehashing a flight or talking about real or imagined flight experiences.

Hangar session—A meeting of a pilot group, regardless of whether it takes place in a hangar or elsewhere.

Hang glider—A winged contraption with or without operable flight controls, having neither engine nor fuselage. The pilot usually hangs under a

SIGNALMAN DIRECTS TOWING

SIGNALMAN'S POSITION

FLAGMAN DIRECTS PILOT TO SIGNALMAN IF TRAFFIC CONDITIONS REQUIRE

ALL CLEAR (O.K.)

START ENGINE

POINT TO ENGINE TO BE STARTED

STOP

LEFT TURN

PULL CHOCKS

COME AHEAD

RIGHT TURN

INSERT CHOCKS

EMERGENCY STOP

NIGHT OPERATION

SLOW DOWN

CUT ENGINES

Hand signals used by line personnel (courtesy of FAA).

pair of fabric wings and controls the flight by shifting the weight of his body.

Hang gliding—The action of flying a hang glider. It is the only aviation activity so far not under the jurisdiction and/or control of the FAA.

HAT—Height above touchdown.

Have the numbers—Expression used by pilots when approaching an airport or getting ready to depart, to inform the tower or ground control-

Cessna Hawk XP (courtesy of Cessna).

ler that they have received the current information with regard to weather, wind, active runway and so on.

Hawk—Short for Skyhawk, a single-engine fixed-gear aircraft manufactured by Cessna Aircraft Company.

Hawker Siddeley Aviation Ltd.—Manufacturers of a variety of aircraft. (Richmond Road, Kingston upon Thames, KT2 5Qs, Surrey England. U.S. representative: British Aerospace, Inc., P.O. Box 17414, Washington, D.C. 20041).

Haze—Restriction of visibility caused by dust, industrial pollution, salt particles in the air or other causes. The sequence report symbol for haze is H.

HDI—Horizontal deviation indicator. See artificial horizon.

Heading—A direction expressed in numbers of degrees from a given reference point, counting clockwise. Headings can be true (based on true north), magnetic (based on magnetic north) or compass headings.

Heading indicator—Directional gyro.

Headset—A combination earphone(s) and microphone which is worn on the head in order to be able to communicate while keeping one's hand free.

HS-125-731 (courtesy of British Aerospace).

British Aerospace (Hawker Siddely) jet aircraft specifications.

BRITISH AEROSPACE MODEL:	HS 125-700
ENGINES	GA/TFE 731-3R (2)
THRUST (pounds each)	3,700
SEATS crew + passengers	2 + 6 to 12
BALANCED FIELD LENGTH, ISA (ft)	6,250
ISA + 20 (ft)	8,000
RATE OF CLIMB (fpm)	3,900 (1 eng. 1,350)
Vr (knots)	137
V2 (knots)	138
SERVICE CEILING (ft)	41,000 (1 eng. 25,000)
Mmo	.78
ALTITUDE CHANGEOVER (FL)	270
Vmo (knots)	320
CRUISE hi speed (knots)	427 (FL 350)
long-range (knots)	390 (FL 370)
RANGE long-range cruise, w. res. (nm)	2,350
FUEL FLOW per engine (pph) hi speed	850
economy	675
RAMP WEIGHT (lbs)	25,000
TAKEOFF/LANDING WEIGHT (lbs)	24,800/22,000
ZERO FUEL WEIGHT (lbs)	16,050
USEFUL LOAD (lbs)	11,600
FUEL CAPACITY (lbs)	9,450
MAX PAYLOAD (lbs)	2,400
FUEL w. Max payload (lbs)	8,800
PAYLOAD w. full fuel (lbs)	1,750
WING LOADING (lbs per sq ft)	70.3
WING AREA (sq ft)	353
LENGTH/HEIGHT/SPAN external (ft)	50.75/17.6/47
LENGTH/HEIGHT/WIDTH cabin (ft)	21.3/5.8/5.9
PRESSURIZATION (psi)	8.4
EXTERIOR NOISE (EPNdB)	87.6/96.3
PRICE (1979 $s)	3,050,000

Head-up display—(HUD)—A means of presenting nav information in a manner which permits the pilot to see both this information and the outside through his windscreen.

Headwind—Wind blowing directly against the line of flight of the aircraft.

Headwind component—The amount of reduction in ground speed resulting from wind blowing toward the aircraft from an angle more or less coinciding with the direction of flight. Even a 90-degree crosswind produces a headwind component.

Heavy aircraft—Aircraft capable of takeoff weights of 300,000 pounds or more, regardless of the weight at which they are operating at any given time.

Height above airport—(HAA)—The height of the MDA above the surface of the airport, as used in connection with the published IFR circling-approach minimums.

Height above landing—(HAL)—The height above a designated helicopter landing area. Used only in helicopter IFR approach procedures.

Height above touchdown—(HAT)—The height of the MDA or DH above the highest elevation of the touchdown zone. It is published on instrument approach charts in connection with straight-in approach minimums.

Helicopter—A rotary-wing aircraft in which both lift and forward speed are produced by an engine driving the rotar, the blades of which tilt in a manner which results in lift as well as thrust.

Helicopter Association of America—(HAA)—See Aviation Association and organizations.

Helio—A family of high-performance STOL single-engine aircraft equipped with automatic leading-edge slats for extra lift at low speeds.

Helipad—The actual touchdown portion of a heliport; or any temporary or permanent installation designed to permit helicopter landings.

Heliport—An area designed to be used for landing, takeoff and loading of helicopters. It may be on land, on water or atop a building.

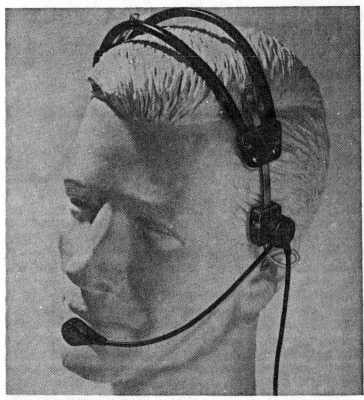

Headset (courtesy of Telex).

Hughes 300 helicopter (courtesy of Hughes).

Helium—A non-flammable gas used as a lift agent in gas balloons and dirigibles.

Hemispheric rules—FARs governing the altitudes at which an aircraft should fly, depending on which half circle of the compass (hemisphere) its magnetic heading falls, in order to achieve a reasonable degree of vertical separation. See *cardinal altitudes*. The hemispheric rules do not apply to aircraft under ATC control.

Hertz—(Hz)—A measurement of the frequency of radio waves equal to one cycle per second. It is named after Heinrich R. Hertz who proved the existence of electro-magnetic waves.

HF—High frequency.

HF transceivers—High frequency com equipment.

HG—Mercury (hydrargyrum.) Used primarily in expressing manifold pressure in inches of mercury (in. Hg).

HGT—Height.

HI—High.

HIALS—High intensity approach light system.

HIGE—Hovering in ground effect. Used with reference to helicopters to give the service ceiling for hovering in ground effect. Given in feet msl.

High—A region of atmospheric pressure surrounded by lower pressure and, in the northern hemisphere, by clockwise winds. Also called anticyclone.

High clouds—Clouds with bases above 20,000 feet.

High frequency—(HF)—Radio frequencies from three to 30 MHz (3,000 to 30,000 cycles per second).

High frequency communications—Com radios using the HF band for long-range communications.

High lift devices—Any devise, generally attached to the wing of an aircraft, used to increase lift, primarily during slow flight, takeoff or

Helio Courier executing a short-field takeoff.

High frequency transceivers comparison chart.

MANUFACTUER	MODEL	PRICE uninstalled	DC	AC 400 Hz	CHANNELS number	Power output watts PEP	Units	Weight lbs.
BRELONIX	SAM-70	1,595	14		5	40	1	5
BRELONIX	SAM-70	1,695	28 14		5	40	1	5
PANTRONICS	PT 10A	1,800	28 14		10	40	2	12
PANTRONICS	DX 10DA	1,950	28 14		10	30	2	15.1
PANTRONICS	DX 10 RA	1,950	28 14		10	40	3	15.5
BRELONIX	SAM-100-5	1,995	28 14		5	100	2	12.5
BRELONIX	SAM-70	1,995	28 14		5	40	2	6.5
BRELONIX	SAM-100-10	2,550	28 14		10	100	2	12.5
SUNAIR	ASB 60	2,645	28 14		6	125	2	12.5
PANTRONICS	SB10	2,750	28 14		10	100	2	19
SUNAIR	ASB 125	2,965	28 14		10	125	2	13
SUNAIR	ASB 100A	3,295	28 14		10	100	3	18
SUNAIR	ASB 130	3,450	28		10	130	3	14
SUNAIR	ASB 500	3,500	28		32,000	100	2	16
SUNAIR	ASB 320	4,395	28		20	130	2	20.25
COLLINS	HF-200	7,480	28		20	100	4	24.5
SUNAIR	ASB 850	9,750	28		280,000	100	3	42.6
COLLINS	618T-3B	18,536	27.5	115	280,000	400	1	
GOULD	AN ARC-98	N.A.	28		280,000	400	4	60.4
COLLINS	718U-5	N.A.	28		280,000	100	2	33

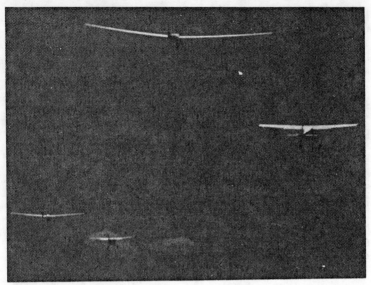

High tow.

landing. Trailing-edge flaps and leading-edge slats are the most frequently used high-lift devised.

High pressure system—See high.

High speed exit—A taxiway angled to the runway in such a way as to permit aircraft to exit the active runway without the need for excessive speed reduction.

High-speed stall—A stall occurring at high airspeeds. An airfoil of a given shape always stalls at a specific angle of attack. An airplane can be made to stall at high speed by a rapid change of the angle of attack, such as might occur in a snaproll. The stall speed of an airplane increases with an increase in wing loading caused by the amount of weight carried in the aircraft or by the increase in wing loading effected by steep turns.

High speed taxiway—See high speed exit.

High speed turnoff—See high speed exit.

High tow—In soaring, the method of towing a glider behind a powered aircraft during which the glider flies above the flight path of the tow plane.

High wing—The position of the wing at or near the top of the fuselage on a monoplane.

Hiller Aviation Division—Manufacturers of piston helicopters. (2075 West Scranton Avenue, Porterville, CA 93257).

HIRL—High intensity runway light system.

HLF—Half.

HLSTO—Hailstones.

HND—Hundred.

Hiller UH-12-E spraying an orange grove (courtesy of Hiller).

Hoffman Electronics—See Gould, Inc.

HOGE—Hovering out of ground effect. Used with reference to helicopters to give the service ceiling when hovering out of ground effect. Given in feet msl.

Holding—Flying a race-track shaped pattern while waiting for further clearance.

Holding fix—A specified location which can be recognized by the pilot either by reference to a ground feature or by using one or several nav aids, and used to establish and maintain a given position of an aircraft while holding.

Hiller Aviation Division utility helicopters specifications.

HILLER AVIATION DIVISION MODEL:	UH12E steel blades	UH12E4	UH12E Soloy conversion
ENGINE	L/VO-540	L/VO-540	AL/250-C2C
hp or shp	305	305	301
SEATS	3	4	3
RATE OF CLIMB max at sea level (fpm)	993	1,290	1,706
vertical (fpm)	230	740	1,463
SERVICE CEILING (ft)	10,300	16,200	NA
HIGE (ft)	7,600	10,800	12,000
HOGE (ft)	3,900	7,200	9,500
MAX SPEED sea level (knots)	65	83	90
NORMAL CRUISE SPEED (knots)	61	78	83
FUEL CAPACITY (lbs)	276	276	308
RANGE w. full fuel (nm)	127	153	305
GROSS WEIGHT (lbs)	3,100	2,800	3,100
EMPTY WEIGHT (lbs)	1,759	1,836	1,650
USEFUL LOAD (lbs)	1,341	964	1,450
HEIGHT/LENGTH/WIDTH external (ft)	9.3/40.7/10	9.3/40.7/10	9.3/40.7/10
MAIN ROTOR number of blades	2	2	2
diameter (ft)	35.4	35.4	35.4
PRICE (1979 $s)	102,500	115,400	159,400

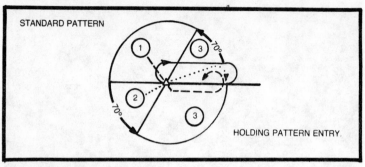

Holding pattern entry (courtesy of FAA).

Holding pattern—A precise pattern, usually elliptical, based on one or several nav aids which an airplane is supposed to fly at an altitude given by ATC while waiting for further clearance. Frequently assigned holding patterns are shown on the low-altitude en-route charts.

Holding pattern entries—Procedures recommended by the FAA (but not mandatory) which are designed to make it relatively easy for a pilot to establish himself in a holding pattern. There are three basic types: Direct entry, parallel entry, and teardrop entry.

Holding procedure—A standard maneuver used in the air and/or on the ground to keep an aircraft within a given geographical area while awaiting further clearance.

Homebuilt—An aircraft on which at least 51 percent of the construction was performed by an amateur builder.

Homing—Flying toward a nav aid.

Homing beacon—A non-directional radio beacon (NDB) which broadcasts a constantly repeating Morse code signal identifying the station. It uses M/LF frequencies and is received by the ADF in the aircraft.

Hood—A contraption worn by flight students during instrument flight instruction to prevent them from being able to see outside the aircraft.

Hop—A short flight from one airport to another nearby.

Horizon—The line at which the earth and the sky appear to be meeting.

Horizon—The reference line on an artificial horizon representing the actual horizon with reference to the attitude of the aircraft. Similar horizon indications are included in the pictorial displays which are part of flight-director systems.

Horizontally opposed engine—A reciprocating engine in which pairs of cylinders are opposite to one another on a horizontal plane with the crankshaft between them.

Horizontal situation indicator—(HSI)—A cockpit display activated by nav receivers and usually including a slaved compass card, which by means of various moving indicators shows the pilot his position relative to certain given nav aids. Most HSIs include a VOR and ADF readout as

A row of homebuilt aircraft at the EAA convention and fly-in at Oshkosh.

Horizontal situation indicator (HSI) (courtesy of Sperry).

Horizontal situation indicators comparison chart.

MANUFACTURER	MODEL	PRICE uninstalled	DC	AC 400 Hz	Vacuum required	panel	remote	autopilot	RMI	Slaved	Units	Weight lbs.	REMARKS
EDO-AIRE/MITCHELL	DG 360A	1,775	14 28					OPT	OPT		1	4.5	
EDO-AIRE/MITCHELL	NSD 360A	2,525	14 28								1	4.5	
NARCO	HSI 100S	2,680	14 28						OPT		1	4	
ARC	IG-832C	3,475	28								1	4.4	
NARCO	HSI 1005	3,495	14 28					OPT			2	5.4	
ELECTRONIQUE AEROSPATIALE	HSI 651	3,500		26 115							1	3.3	
KING	KCS 55A	3,515	14 28								4	9.4	
COLLINS	331A-3G	3,717	28	26							N.A.		STANDARD ARINC INPUTS
EDO-AIRE/MITCHELL	NSD 360A(S)	3,800	14 28					OPT	OPT		3	5.1	
EDO-AIRE/MITCHELL	NSD 360A(S) /4	4,180	14 28								3	5.1	
ARC	IG-832 A	4,670	28								3	5.9	
BENDIX AVIONICS	HDS-830	4,900	14 28	26							6	7.5	
KING	KPI 552	5,200	28	115							4	10	REQUIRES KGS 105 OR ANY STANDARD ARINC SLAVED DIRECTIONAL GYRO SYSTEM
BENDIX AVIONICS	HSD-880	5,670	14 28								6	11.1	INCLUDES RMI IN VOR AND ADF MODES. (NO EXTERNAL RMI REQUIRED)
KING	KPI 553	6,565	28	115							4	10.5	REQUIRES KGS 105 OR ANY STANDARD ARINC SLAVED DIRECTIONAL GYRO SYSTEM
COLLINS	PN-101	6,980	28								N.A.		STANDARD ARINC INPUTS
COLLINS	331A-6R	7,196	28	115							N.A.		STANDARD ARINC INPUTS
SPERRY	RN-Z00	8,670	28	115							3	10.8	AVAILABLE FOR 28 V (NO 115 AC) AT $540 EXTRA
COLLINS	331A-9G	9,885	28	115							N.A.		STANDARD ARINC INPUTS, 331A-9G IS A PART OF FD-109 FLIGHT DIRECTOR

well as a heading information and glide-scope indication, all in one instrument. HSIs are a standard component of flight-director systems, but can be installed and used without a flight director.

Horizontal stabilizer—The fixed horizontal section of the empennage to which the elevators are attached. Some aircraft are equipped with so-called stabilators, in which case the entire horizontal stabilizer can be moved by the pilot, acting as an elevator in its entirety.

Horsepower—A unit of energy equal to the power needed to raise 550 pounds one foot in one seccond. Generally used in measuring engine output.

Hotel—In aviation-radio phraseology the term used for the letter H.

Hovering—The ability of a helicopter to retain a given altitude without forward, rearward or lateral movement.

How do you hear me—Radio phraseology for, "Did you understand what I said?"

How do you read (me)—Same as *how do you hear me*. Often abbreviated to: Do you read?

hp—Horsepower.

HR—Hear.

HR—Here.

Hawker Siddley HS 125 (courtesy of Hawker Siddeley).

hr—hour.

HS 125—A family of corporate jet aircraft manufactured by Hawker Siddeley Aviation in England. Also intermittently referred to as DH 125 (DeHavilland) and BH 125 (Beech-Hawker).

HSI—Horizontal situation indicator.

HST—Hawaiian Standard time.

HT Instruments—Manufacturers of certain specialized avionics equipment. (4121 Redwood Avenue, Los Angeles, CA 90066).

Hub—A major airport.

HUD—Head-up display.

Hughes Helicopters—Manufacturers of a variety of piston and turbine helicopters. (Centinela and Teale Streets, Culver City, CA 90230).

Hughes 500 D helicopter (courtesy of Hughes).

177

Humidity—The amount of water vapor in the air, expressed either in terms of absolute or relative humidity.

Hunting and Dragging—See dragging.

HURCN—Hurricane.

Hurricane—A tropical cyclone with wind speeds of 64 knots or more, usually originating in the vicinity of the West Indies.

Hustler—A unique aircraft developed by American Jet Industries and now marketed by Gulfstream-American Corporation. It comes in three versions, one as a single-engine turboprop with the turboprop in the front, driving a nose-mounted propeller. A second verson adds a Pratt and Whitney fanjet engine in the rear of the aircraft. A third version uses a tail mounted small auxiliary fanjet engine developed by Williams Research.

HVY—Heavy.

HWY—Highway.

HWVR—However.

Hydraulics—A branch of science that deals with practical applications of liquids in motion, such as the transmission of energy by compressing liquids.

Hydraulic system—A system of valves, lines and pipes in the aircraft which uses the pressure of fluids, usually some type of oil, to move a structure or structures. It is based on the scientific fact that pressure applied to a fluid will pass through it and be concentrated in a small area of least resistance. Thus small pressures on the controls will turn into large

Hughes Helicopters specifications.

HUGHES HELICOPTERS MODEL:	300C (piston)	500 D (turbine)
ENGINE	L/HIO-360-D1A	AL/250-C20B
hp or shp	225	420
SEATS	3	5 to 7
RATE OF CLIMB max at sea level (fpm)	1,305	1,900
vertical	NA	900
SERVICE CEILING (ft)	14,900	15,000
HIGE (ft)	5,900	8,500
HOGE (ft)	2,750	7,500
MAX SPEED sea level (knots)	91	151
CRUISE SPEED (knots)	83	130
FUEL CAPACITY (lbs)	180 or 294	429
RANGE w. full fuel	166 or 274	221
GROSS WEIGHT (lbs)	2,050	3,550
EMPTY WEIGHT (lbs)	1.046	1,260
USEFUL LOAD (lbs)	1,004	2,290
PAYLOAD w. full fuel (lbs)	824 or 710	1,861
LENGTH/HEIGHT/WIDTH external (ft)	30.83/8.75/6.54	30.5/8.9/6.8
MAIN ROTOR number of blades	3	5
diameter (ft)	26.84	26.5
TAIL ROTOR number of blades	2	2
diameter (ft)	4.25	5.5
PRICE (1979 $s)	76,000	227,000

Hustler. See the chart listing Hustler 500 specifications on page 120. (courtesy American Jet Ind.).

pressures capable of moving the control surfaces, landing gear or other systems. Hydraulic systems must always maintain a given amount of fluid under given pressure.

Hydrogen—A highly flammable gas used as a lifting agent in gas balloons and dirigibles. Being vastly less expensive than the safer (non-flammable) helium, it continues to be used. The inability of Germany to obtain and pay for helium in the operation of its Zeppelins, resulted in the use of hydrogen and, in the final analysis, the Hindenburg disaster.

Hygrometer—An instrument that measures humidity.

Hypoxia—A deficiency of oxygen reaching the tissues of the body. It occurs when flying at high altitudes without supplemental oxygen, initially producing a misleading feeling of euphoria, then inhibiting the pilot from being able to function efficiently and eventually resulting in unconsciousness. Initial onset of hypoxia can be detected by the color of the fingernails. They tend to turn blueish purple.

Hz—Hertz.

I—India (phonetic alphabet).

IAF—Initial approach fix.

IAP—Instrument approach procedure.

IAS—Indicated airspeed.

IATA—International Air Transport Association.

IC—Ice crystals (in sequence reports).

ICAO—International Civil Aviation Association.

Ice—One of the two most dangerous phenomena in aviation. The other: Thunderstorms.

Ice fog—Ice fog consists of tiny ice crystals, in fact, frozen water vapor. It tends to be most prevalent in the vicinity of populated areas (and airports) where man-made pollutants are the nuclei around which the ice crystals form.

Ice protection systems—A variety of systems is employed in aviation to protect aircraft from the effects of ice accumulation: pulsating de-icing boots on the leading edges of airfoils; a means of spraying alcohol onto the propeller blades in flight; heated windshields and heating system in the leading eges of wings and/or other airfoils. And, to counteract carburetor ice, the carburetor-heat system.

ICG—Icing.

ICGIC—Icing in clouds.

ICGICIP—Icing in clouds and precipitation.

ICGIP—Icing in precipitation.

Icing conditions—Weather conditions conducive to depositing ice on an aircraft in flight. Usually characterized by visible moisture in the air and temperatures around the freezing level. Pilots encountering icing conditions are urged to issue PIREPs immediately. When ice buildup occurs, the best evasive action is to climb or descend to either colder or warmer air.

IDC—Intercontinental Dynamics Corporation.

Ident—Phrase used by ATC to ask the pilot to push the ident button on his transponder for positive identification on the radar scope.

IDENT—Identification.

Ident feature—One of the features of the ATC radar systems (ATCRBS) used to instantly distinguish one radar return from another.

Idle—To run the engine(s) at low speed, preferably at an rpm setting recommended by the manufacturer.

IDO—International District Office (FAA).

IF—Ice fog (in sequence reports).

IFF—Identification, friend or foe.

If feasible reduce (increase) speed to (speed)—An ATC phrase used to ask the pilot to make an adjustment in his speed, usually for the purpose of maintaining safe separation between aircraft.

IFIM—International Flight Information Manual.

If no transmission received for (minutes) or by (time)—An ATC phrase used during or prior to instrument approaches to tell the pilot what action to take in the event of communication failure.

IFR—Instrument flight rules.

IFR—Instrument Flight Research Corporation.

IFR aircraft—An aircraft operating under ATC control, or an aircraft appropriately equipped to operate in IFR conditions.

IFR conditions—Weather conditions below the minimums required for VFR flight.

IFR flight—Flight in below-VFR conditions, under ATC control when in controlled airspace.

IFR minimums—Minimums in terms of ceiling and/or visibility under which IFR departures and/or approaches are permissible. They are listed on the instrument approach charts. There are no IFR departure minimums for flights carrying neither passengers nor freight for revenue.

IFR over the top—The operation of an aircraft when cleared by ATC to operate in and maintain VFR conditions on top (above clouds).

IFR reserves—A 45-minute reserve in terms of fuel which must be aboard the aircraft when reaching its destination on an IFR flight.

IFSS—International Flight Service Station.

Ignition check—Part of the preflight procedure during which the pilot of a piston-engine aircraft checks the action of the spark plugs by alternately switching from one set to the other.

Ignition switch—The switch which activates the ignition system of an aircraft.

Ignition system—A system of dual spark plugs, generally powered by magnetos, which provides the spark for igniting the fuel-air mixture in the combustion chamber. It is not part of the electrical system of an aircraft.

IHP—Indicated horsepower.

ILS—Instrument landing system.

ILS categories—There are five categories of ILS approaches; all except one require special certification of the pilot and special instrumentation in the aircraft over and above that required for a standard ILS approach.

Category I (CAT I)—The basic ILS approach, calling for an HAT of not less than 200 feet and an RDR of not less than 1,800 feet.

Category II (CAT II)—Calls for an HAT of not less than 100 feet and an RVR of not less than 1,200 feet.

Category III is divided into three sub-categories:

CAT IIIA —Calls for no decision height minimum but an RVR of not less than 700 feet.

CAT IIIB—No DH and an RVR of not less than 150 feet.

CAT IIIC—No DH and no RVR minimum. In other words, zero-zero conditions.

IM—Inner marker.

IMC—Instrument meteorological conditions.

IMD—Immediate.

IMDTLY—Immediately.

IMEP—Indicated mean effective pressure.

Immediate—Used by ATC in the context of: Cleared for immediate takeoff; usually because another aircraft is on final.

Immediately—Used by ATC to inform the pilot that he must comply with an instruction without delay in order to avoid a possibly critical situation.

Immelmann—The Immelmann is an aerobatic maneuver consisting of the first half of a loop followed by a half roll to level flight. It derived its name from the German WW I ace who is said to have first perfected the maneuver for the purpose of rapidly gaining altitude while, at the same time, reversing his direction of flight.

Impeller—A vital part of a centrifugal turbine engine. Air enters the impeller through inlet guide vanes. The impeller rotates at high speed, compressing the air by centrifugal action.

IMPT—Important.

in—Inch; inches.

Inbound—Flying toward an airport or nav aid.

Inches of mercury—(in Hg)—A unit of measurement of atmospheric pressure, indicating the height in inches to which a column of mercury (Hg) will rise in a glass tube in response to the weight of the atmosphere exerting pressure on a bowl of mercury at the base of the tube.

Incident—A minor accident or failure or malfunction of an aircraft component or system. Certain incidents require notification of the National Transportation Safety Board (NTSB). Rules covering incidents (and accidents) can be found in FAR Part 430.

INCR—Increase.

Increase speed to (speed)—An ATC instruction to the pilot to increase the speed of his aircraft, usually in order to maintain safe separation between aircraft.

INDC—Indicate.

INDEF—Indefinite; indefinitely.

India—In aviation-radio phraseology the term used for the letter I.

Indicated airspeed—The airspeed of an aircraft as displayed on its airspeed indicator (IAS).

Indicated altitude—Height above sea level as shown on the altimeter when it has been set to the appropriate barometric pressure corrected to msl.

Indicated horsepower—The actual power produced in the engine combustion chamber.

Indicated mean effective pressure—The average of combustion pressures exerted on the combustion chamber.

Induced drag—Drag created through the process of producing lift.

Induction system—The air scoop and related plumbing.

Induction system icing—Under certain atmospheric conditions ice may form and eventually block the induction system. For this reason most aircraft are equipped with an alternate air source which, when used, takes air from the cabin or other area permanently protected from ice. Using alternate air usually results in a reduction of power.

Inertia—The property of matter by which it remains at rest or in uniform motion in the same straight line unless acted upon by some external force.

Inertial navigation—Navigating by using the principle of inertia. A gyroscopic instrument on the aircraft senses acceleration, deceleration and changes in direction and, feeding this information into a computer, tells the pilot his position at any given time, assuming the computer has been told the departure point in terms of latitude and longitude.

Inertial navigation system—The hardware required for inertial navigation. Prices for these systems run in the neighborhood of $100,000, and as a result such systems are only found on airliners and corporate jets used for intercontinental flights.

Inflation—The act of feeding hot air or gas into a balloon or blimp.

Inflation sleeve—The sleeve attached to a gas balloon through which gas is fed into it.

INFO—Information.

Information request—(INREQ)—A request for information originated by an FSS in an effort to locate an overdue or missing aircraft.

in Hg—Inches of mercury.

Initial approach fix—(IAF)—A fix shown on instrument approach charts and representing the start of the first segment of the instrument approach procedure.

Initial approach segment—The first segment of an instrument approach, starting at the IAF.

Injection—Fuel injection.

Inner ear—The portion of the ear providing a sense of balance. See also Eustachian tubes.

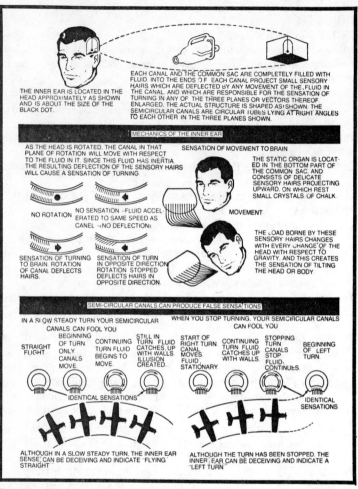

Function of the inner ear (courtesy of FAA).

Inner marker—A marker beacon used primarily in CAT II approaches. It is located more or less half way between the middle marker (MM) and the runway threshold.

INOP—Inoperative.

INREQ—Information request.

INS—Inertial navigation system.

Inspection of aircraft—The FARs require periodic inspection of aircraft by an inspector authorized by the FAA. Such inspections must be logged in the aircraft logbook and initialed by the inspector.

INST—Instrument.

Installation error—Error in the reading of an airspeed indicator caused by the difference between the actual and theoretical comparison between pitot and static pressure. Calibrated airspeed accounts for this error and makes the appropriate correction.

Instrument approach—Any approach made by reference to instruments and ground-based navigation aids or communication systems.

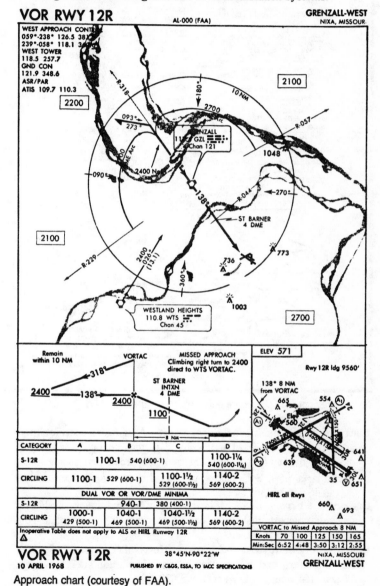

Approach chart (courtesy of FAA).

185

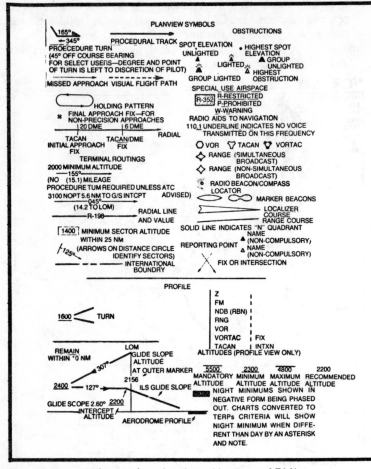

Instrument approach procedure chart legend (courtesy of FAA).

Instrument approach procedures chart—A chart of a specific airport area, showing the airport and facility diagrams in detail, plus giving all directions needed for making an instrument approach and landing at that particular airport. Such charts are issued by the government and by Jeppesen-Sanderson.

Instrument error—Error in the reading of an instrument, such as a magnetic compass, that is inherent in that particular instrument, and the degree of which is known.

Instrument Flight Research Corporation—Manufacturers of encoding altimeters and other aviation instrumentation. (2716 George Washington Boulevard, Wichita, KS 67210.)

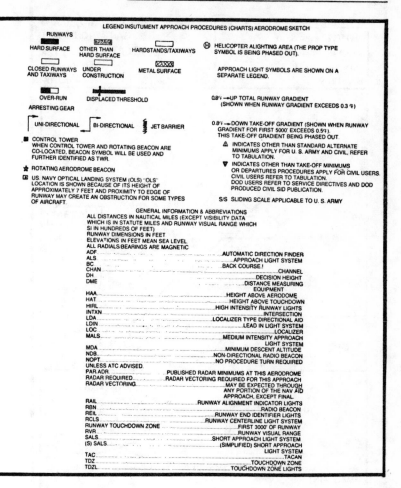

LEGEND INSTUMENT APPROACH PROCEDURES (CHARTS) AERODROME SKETCH

RUNWAYS

HARD SURFACE   OTHER THAN HARD SURFACE   HARDSTANDS/TAXIWAYS

CLOSED RUNWAYS AND TAXIWAYS   UNDER CONSTRUCTION   METAL SURFACE

OVER-RUN   DISPLACED THRESHOLD

ARRESTING GEAR

UNI-DIRECTIONAL   BI-DIRECTIONAL   JET BARRIER

■ CONTROL TOWER
WHEN CONTROL TOWER AND ROTATING BEACON ARE CO-LOCATED, BEACON SYMBOL WILL BE USED AND FURTHER IDENTIFIED AS TWR.

★ ROTATING AERODROME BEACON

⊡ US. NAVY OPTICAL LANDING SYSTEM (OLS) "OLS" LOCATION IS SHOWN BECAUSE OF ITS HEIGHT OF APPROXIMATELY 7 FEET AND PROXIMITY TO EDGE OF RUNWAY MAY CREATE AN OBSTRUCTION FOR SOME TYPES OF AIRCRAFT.

Ⓗ HELICOPTER ALIGHTING AREA (THE PROP TYPE SYMBOL IS BEING PHASED OUT).

APPROACH LIGHT SYMBOLS ARE SHOWN ON A SEPARATE LEGEND.

0.8% →UP TOTAL RUNWAY GRADIENT
(SHOWN WHEN RUNWAY GRADIENT EXCEEDS 0.3%)

0.8% → DOWN TAKE-OFF GRADIENT (SHOWN WHEN RUNWAY GRADIENT FOR FIRST 5000' EXCEEDS 0.5%).
THIS TAKE-OFF GRADIENT BEING PHASED OUT.

△ INDICATES OTHER THAN STANDARD ALTERNATE MINIMUMS APPLY FOR U. S. ARMY AND CIVIL, REFER TO TABULATION.

▽ INDICATES OTHER THAN TAKE-OFF MINIMUMS OR DEPARTURES PROCEDURES APPLY FOR CIVIL USERS. CIVIL USERS REFER TO TABULATION.
DOD USERS REFER TO SERVICE DIRECTIVES AND DOD PRODUCED CIVIL SID PUBLICATION.

S/S SLIDING SCALE APPLICABLE TO U. S. ARMY

GENERAL INFORMATION & ABBREVIATIONS
ALL DISTANCES IN NAUTICAL MILES (EXCEPT VISIBILITY DATA WHICH IS IN STATUTE MILES AND RUNWAY VISUAL RANGE WHICH SI IN HUNDREDS OF FEET)
RUNWAY DIMENSIONS IN FEET
ELEVATIONS IN FEET MEAN SEA LEVEL
ALL RADIALS/BEARINGS ARE MAGNETIC

ADF	AUTOMATIC DIRECTION FINDER
ALS	APPROACH LIGHT SYSTEM
BC	BACK COURSE.!
CHAN	CHANNEL
DH	DECISION HEIGHT
DME	DISTANCE MEASURING EQUIPMENT
HAA	HEIGHT ABOVE AERODOME
HAT	HEIGHT ABOVE TOUCHDOWN
HIRL	HIGH INTENSITY RUNWAY LIGHTS
INTXN	INTERSECTION
LDA	LOCALIZER TYPE DIRECTIONAL AID
LDIN	LEAD IN LIGHT SYSTEM
LOC	LOCALIZER
MALS	MEDIUM INTENSITY APPROACH LIGHT SYSTEM
MDA	MINIMUM DESCENT ALTITUDE
NDB	NON-DIRECTIONAL RADIO BEACON
NOPT	NO PROCEDURE TURN REQUIRED UNLESS ATC ADVISED.
PAR.ADR.	PUBLISHED RADAR MINIMUMS AT THIS AERODROME
RADAR REQUIRED	RADAR VECTORING REQUIRED FOR THIS APPROACH
RADAR VECTORING	MAY BE EXPECTED THROUGH ANY PORTION OF THE NAV AID APPROACH, EXCEPT FINAL
RAIL	RUNWAY ALIGNMENT INDICATOR LIGHTS
RBN	RADIO BEACON
REIL	RUNWAY END IDENTIFIER LIGHTS
RCLS	RUNWAY CENTERLINE LIGHT SYSTEM
RUNWAY TOUCHDOWN ZONE	FIRST 3000' OF RUNWAY
RVR	RUNWAY VISUAL RANGE
SALS	SHORT APPROACH LIGHT SYSTEM
(S) SALS	(SIMPLIFIED) SHORT APPROACH LIGHT SYSTEM
TAC	TACAN
TDZ	TOUCHDOWN ZONE
TDZL	TOUCHDOWN ZONE LIGHTS

Instrument flight rules—Rules governing the procedures for flight under IFR conditions and/or under the direction of ATC. (FAR Part 61 and 91).

Instrument flying—Flying by reference to instruments, ground-based nav aids and/or communication with ATC.

Instrument landing system—(ILS)—An electronic system of landing aids which provide precise directional and altitude information to the pilot through instrument display(s) in the cockpit. Directional information is supplied by the localizer, altitude information by the glide slope. ILS minimums generally are a 200-foot ceiling and a half mile visibility. An ILS approach is a precision approach.

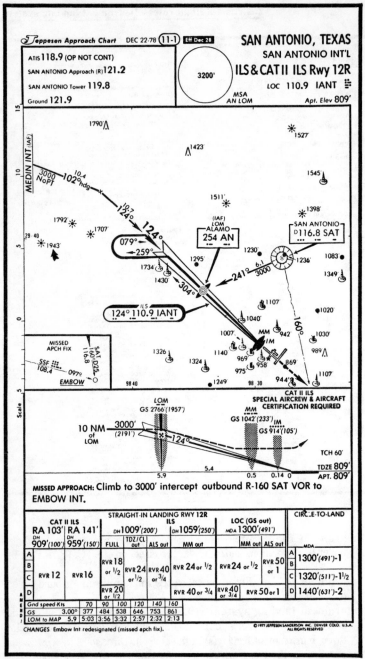

Instrument approach procedure chart issued by Jeppesen-Sanderson (courtesy of Jeppesen Sanderson, Inc.).

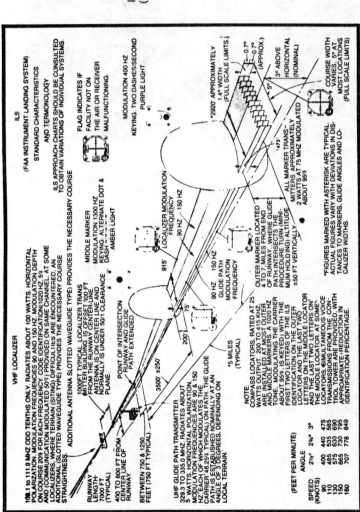

Instrument landing system
(courtesy of FAA).

Instrument panel (courtesy of Cessna).

Instrument meteorological conditions—Weather conditions in terms of ceiling, visibility, and cloud types below those which constitute VFR minimums.

Instrument panel—The panel housing the instrument displays (though not necessarily the complete instruments) located in front of the pilot and copilot.

Instrument rating—The rating which a pilot must have in order to fly approved IFR flights. Obtaining an instrument rating requires a minimum of 250 hours as pilot in command, considerable flight training and a written as well as a flight test.

Instrument runway—A runway equipped with the appropriate electronic and visual aids for which an instrument approach, either precision or non-precision, has been approved and published.

Instrument ticket—Instrument rating.

Insurance—Aviation related insurance (liability, hull, etc.) is a very specialized field and should be discussed with, and purchased from, a person or company specializing in aviation insurance.

INT—Intersection.

Integrated autopilot/flight-director systems—Flight control systems in which the autopilot is coupled to the flight director and takes its commands from the flight director. These systems can be manually overridden at the pilot's discretion.

Intercept—The action of flying toward and then aligning the aircraft with a VOR radial or bearing.

Interception—Intercept.

Intercontinental Dynamics Corporation—Manufacturers of a wide variety of aircraft instruments. (P.O. Box 81, Englewood, NJ 07631.)

Interiors—While most general-aviation aircrafts are manufactured and delivered with standard interiors, many of the higher priced corporate aircraft are ordered with custom interiors, and there are a number of expert companies in various parts of the country which specialize in the design, manufacture and installation of custom interiors.

Intermediate approach segment—The segment of an instrument approach between the initial approach segment and the final approach.

Internal combustion engine—Reciprocating engine.

International airport—An airport of entry offering customs service; a landing-rights airport where permission to land must be obtained in advance from customs authorities; airports designated under the convention on international civil aviation as an airport to be used by international air transport and general aviation flights.

International Aviation Theft Bureau—A clearing house for information leading to the recovery of stolen aircraft and aircraft instruments. (7315 Wisconsin Avenue, Washington, D.C. 20014.)

International Civil Aviation Organization—An agency of the United Nations concerned with matters related to international civil air transport.

International Flying Farmers—See *aviation organizations*.

Interrogator—The signal sent by a radar beacon to be re-transmitted by a transponder.

Intersecting runways—Two or more runways which cross or meet one another.

Intersection—A point at which the signals from two or more nav aids cross, usually given a name related to some nearby geographic location or landmark. Shown on charts by open triangles or crossed arrows.

Intersection departure—A takeoff from the pilot along a runway where that runway intersects a taxiway or another runway. Intersection departures may be suggested by tower controllers to expedite traffic but may be refused by the pilot if he prefers to have the entire length of the runway available for takeoff.

Intersection takeoff—Intersection departure.

INTL—International.

INTMT—Intermittent.

INTR—Interior.

INTS—Intense.

INTSFY—Intensify.

INTSV—Intensive.

INTXN—Intersection.

Inversion—A fairly shallow layer of the atmosphere in which the lapse rate is reversed and temperature increases instead of decreasing with altitude. It is usually caused by cooling of the air at the surface without extensive vertical mixing. In an inversion the air is usually smooth, but it tends to hold restrictions to visibility, such as smog, close to the ground.

Inverted attitude recovery—The process of transition from inverted to normal flight, accomplished by rolling toward the bank index pointer on the artificial horizon.

Inverted flight—Flying upside down; an aerobatic maneuver.

Inverted outside loop—It begins from normal level flight, followed by a push-over as for a steep dive. This stick position is maintained until the loop is completed and level flight regained.

Inverted spin—A relatively unpleasant maneuver resulting in considerable negative G forces. It is accomplished by holding the yoke full forward as the aircraft stalls in an inverted or extreme nose-high position. To maintain the spin the yoke must continue to be held forward. Releasing the yoke will usually convert the maneuver into a normal spin.

Inverter—An electrical instrument which transforms DC current of a given voltage into AC current, usually of a different voltage.

Ionosphere—A region of strongly ionized particles in the atmosphere, about 35 to 300 miles above the surface of the earth. It reflects radio waves around the curvature of the earth.

IOVC—In the overcast.

IPV—Improve.

IR—Ice on the runway.

ISA—Standard atmospheric conditions. The abbreviation stands for ICAO Standard Atmosphere.

I say again—I am about to repeat what I said.

Isobar—An imaginary line or a line on a chart connecting places on the surface of the earth where barometric pressure, reduced to sea level, is the same at a given time or for a given period.

Israel Aircraft Industries Westwind at Reading Air Show.

Israel Aircraft Industries jet aircraft specifications.

ISRAEL AIRCRAFT INDUSTRIES MODEL:	WESTWIND I
ENGINES	GA/TFE 731-3-1G (2)
THRUST each (pounds)	3,700
TBO	on condition
SEATS crew + passengers	2 + 7 to 10
BALANCED FIELD LENGTH ISA (ft)	4,950
ISA+20 (ft)	7,650
RATE OF CLIMB (fpm)	4,000 (1 eng. 1,100)
Vr (knots)	134
V2 (knots)	141
SERVICE CEILING (ft)	45,000
Mmo	.765
ALTITUDE CHANGEOVER (FL)	194
Vmo (knots)	360
CRUISE hi speed (knots)	424
economy (knots)	401
RANGE hi speed, max fuel & payload (nm)	2,210
economy, max fuel & payload (nm)	2,440
FUEL FLOW per engine, hi speed (pph)	2,156
economy (pph)	1,896
RAMP WEIGHT (lbs)	23,000
TAKEOFF/LANDING WEIGHT (lbs)	22,850/19,000
ZERO FUEL WEIGHT (lbs)	16,000
USEFUL LOAD (lbs)	10,214
FUEL CAPACITY (lbs)	8,710
MAX PAYLOAD (lbs)	2,410
PAYLOAD w. max fuel (lbs)	1,104
FUEL w. max payload (lbs)	7,404
WING LOADING (lbs per sq ft)	74.1
WING AREA (sq ft)	308.3
LENGTH/HEIGHT/SPAN external (ft)	52.3/15.8/44.8
LENGTH/HEIGHT/WIDTH cabin (ft)	15.3/4.9/4.8
PRESSURIZATION (psi)	8.8/9

Isogenic lines—Lines of identical easterly or westerly magnetic variation, shown on aviation charts with the number of degrees of the variation.

Israel Aircraft Industries—Manufacturers of the Westwind family of corporate jets. (Ben Gurion International Airport, Lod, Israel. U.S. representative: Atlantic Aviation, P.O. Box 1709, Wilmington, DE 19899.)

ITT—Intake turbine temperature.

J—Jet route structure.

J—Juliett (phonetic alphabet).

Jamming—Interference, either electronic or mechanical, which disrupts radio transmissions or the ability of radar to display the desired targets.

Jato—Jet assisted takeoff. Small rocket engines attached to aircraft and used to effect extremely steep short-field takeoffs. Used primarily by the military and jettisoned after use.

Jeppesen charts—Radio-facility and instrument-approach charts produced and marketed by Jeppesen-Sanderson.

Jeppesen-Sanderson, Inc.—Producers of aviation charts and audio-visual flight-training materials. (8025 East 40th Avenue, Denver, CO 80207.)

Jet advisory area—Certain areas along some jet routes designated primarily to provide air-carrier jets and certain other aircraft with flight following, traffic information and vectoring service.

Jet blast—Jet engine exhaust. Jet blast can be dangerous to light aircraft positioned behind a jet on the ground. It should not be confused with wake turbulence.

Jet Electronics and Technology, Inc.—Manufacturers of RNAV systems and other avionics. (5353 52nd Street, Grand Rapids, MI 49508.)

Jet engines—Turbine engines.

Jet Ranger—A family of single-engine turbine helicopters manufactured by Bell Helicopter Textron.

Jet route—Airways to be used by aircraft flying at altitudes above FL 180 (18,000 feet) and up to and including FL 450 (45,000 feet msl), designated by the letter J plus a number, such as J-107.

Jet route structure—The network of jet routes.

Bell Jet Ranger (courtesy of Bell).

JetStar—A four-engine corporate jet manufactured by Lockheed. It was modified by Garrett AiResearch with TFE 731 fanjet engines. The modified JetStars are known as 731-JetStars and the later fanjet version manufactured by Lockheed is known as the JetStar II.

Jet stream—A narrow migrating stream of winds with velocities ranging from 50 knots up to 250 knots and occasionally more. It is usually located somewhere between 20 degrees and 70 degrees north latitude at altitudes between 20,000 and 40,000 feet, though it has been known to occasionally dip down to 15,000 feet and below. Flow is west to east.

Jettisoning external stores—The release of tiptanks, jato engines or other hardware attached to the outside of the aircraft while in flight.

Joints—Any juncture of two or more pieces of metal or other materials used in the construction of the air frame.

Lockheed JetStar II (courtesy of Lockheed).

Lockheed Goergia Company jet aircraft specifications.

LOCKHEED GEORGIA COMPANY MODEL: ▪	JETSTAR II
Engines	GA/TFE 731-3 (4)
Thrust (lbs) each	3,700
TBO	progressive, modular
Seats crew + passengers	2 + 10
Balanced field length, ISA (ft)	6,525
ISA + 20 (ft)	8,425
Rate of climb (fpm) all engines	4,100
1 engine out	2,250
Vr (knots)	136
V2 (knots)	151
Service ceiling (ft) all engines	43,000
1 engine out	28,000
Mmo	.82
Altitude changeover (fl)	224
Vmo (knots)	350
Cruise hi speed at fl 300	475 (38,000 lbs)
long range at fl 370	422 (38,000 lbs)
Range hi speed, max fuel, fl 350 (nm)	2,292
max payload, fl 350 (nm)	2,175
long range, max fuel, fl 370 (nm)	2,770
max payload, fl 370 (nm)	2,603
Fuel flow per eng. hi speed (pph)	830
long range (pph)	598
Ramp weight (lbs)	44,500/36,000
Zero fuel weight (lbs)	27,500
Useful load (lbs)	20,150
Max payload (lbs)	2,750
Fuel w. max payload (lbs)	17,250
Payload with full fuel (lbs)	2,004
Wing loading (lbs per sq ft)	82
Wing area ((sq ft)	542.5
Length/height/span external (ft)	60.4/20.4/54.4
Length/height/width cabin (ft)	28.1/6.1/6.2
Pressurization (psi)	8.9
Exterior noise (epndb)	93.1/87.7/97.4
Price (1979 $s)	NA
Standard equipment: autopilot; de-icing; thrust regersers.	

Joint use restricted area—Restricted areas within which flight may be authorized by the FAA when the area is not being used by the controlling agency.

Joy stick—Jargon for control stick, control wheel, yoke.

JTSTR—Jet stream.

Jug—Slang meaning cylinder of a reciprocating engine.

Juliett—In aviation-radio phraseology the term used for the letter J.

Jumbo jet—Very large aircraft, such as the Boeing 747, the McDonnel-Douglas DC-10, the Lockheed L-1011 and C-5A.

k—Cold (as a designation of an air mass).

K—Kilo (phonetic alphabet).

k—Kilo (1,000 of any measure).

K—Kilometer.

K—KiloHertz.

K—Smoke (in sequence reports).

Katabatic wind—The downslope or drainage wind caused by radiation cooling of a shaded slope.

kg—Kilogram (1,000 grams).

kHz—KiloHertz (1,000 Hertz; electromagnetic waves propagating at 1,000 cycles per second).

Kilo—In aviation-radio phraseology the term used for the letter K.

Kilogram—(kg)—1,000 grams.

KiloHertz—(kHz)—1,000 Hertz; radio-magnetic waves propagating at 1,000 cycles per second.

Kilometer—(km)—3,281 feet; 5/8 of a statute mile. See conversion charts.

Kilowatt—(kw)—1,000 watts.

Kinesthetic sense—The sense of position and balance which detects and estimates motion without reference to vision or hearing. It tends to become unreliable during maneuvers in IFR conditions.

King Air—A family of turboprop aircraft manufactured by Beech Aircraft Corporation.

King Radio Corporation—Manufacturers of a wide variety of high quality avionics equipment. (400 North Rogers Road, Olathe, KS 66061.)

Kitchen gas—Ordinary natural gas used occasionally as a lifting agent in gas balloons. It is highly flammable and less effective than hydrogen, but also less expensive.

km—Kilometer; 1,000 meters.

kmh—Kilometers per hour.

Knot—Nautical miles per hour. One knot equals one nautical mile or 1.15 statute miles per hour.

Known traffic—Aircraft, whether IFR or VFR, whose position, altitude and intentions are known to ATC.

Koch chart—A chart showing the percentage increase in the required takeoff distance and the reduction in climb capability caused by an increase in density altitude (airport elevation and temperature).

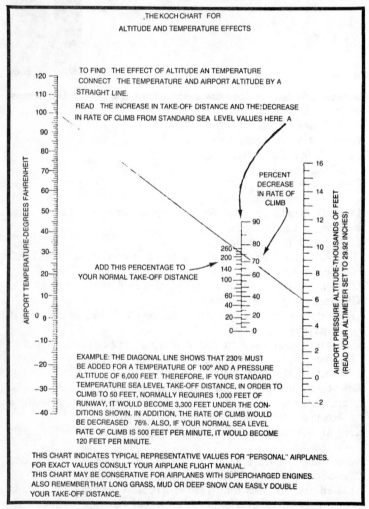

Koch chart for computing takeoff distance and rate of climb (courtesy of FAA).

Beechcraft King Airs (courtesy of Beech).

Kollsman Instrument Company—Manufacturers of a wide variety of flight instruments. (Daniel Webster Highway South, Merrimack, NH 03054.)

Kollsman window—The window on an altimeter displaying the barometric pressure to which the altimeter is set.

KS Avionics—Manufacturers of EGTs and other aircraft instruments. (25216 Cypress Avenue, Hayward, CA 94544.)

kt—Knot.

kw—Kilowatt; 1,000 watts.

kwh—Kilowatt hour.

L—Drizzle (sequence reports).

L—Lights.

L—Lima (pronounced leemah) (phonetic alphabet).

Lake Aircraft—Manufacturers of amphibian aircraft and seaplanes. (P.O. Box 399, Tomball, TX 77375.)

Lake, Island and Mountain Reporting Service—A special service provided by FSSs in certain areas.

Lama—A high-performance utility turbine helicopter manufactured by Aerospatiale in France.

Lambert conformal conic projection—A relatively accurate means of representing the curved surface of the globe on the flat surface of a chart or map.

Laminar airfoil—An airfoil shaped in a way to maintain laminar flow over much of its upper surface.

Laminar flow—Air flowing smoothly over and adhering to the surface of an airfoil.

Laminated—Bonded by means of adhesive agents rather than rivets.

Lance—A high-performance single-engine aircraft of the Cherokee family, produced by Piper Aircraft Corporation.

Land breeze—The movement of air from land to sea at night. It reverses to a sea breeze in the daytime.

Landing—The act of bringing the airplane down to the surface of the earth (or water). Technically, it starts at DH or MDA and includes flare out, touchdown and rollout.

Landing area—Any area of land or water intended to be used for the landing of aircraft, regardless of whether such an area includes actual runways, service facilities and the like.

Lake, Island and Mountain Reporting Service (courtesy of FAA).

Aerospatiale Lama (courtesy of Aerospatiale).

Landing direction indicator—Tetrahedron.

Landing distance—The horizontal distance required to bring an aircraft to a complete stop. It may be measured from the point of touchdown or from an altitude of 50 feet prior to touchdown. It depends on many variables including aircraft weight, wind, runway-surface conditions, touchdown speed, use of flaps, etc. Aircraft flight manuals contain landing-direction charts for standard conditions.

Landing gear—The entire mechanism including hinges, supports, wheels, tires (or floats) necessary in order to support the aircraft on the ground (or water). Landing gears may be fixed or retractable.

Landing gear doors—Doors which cover the gear wells when the landing gear is retracted.

Landing-gear-extended speed—The maximum speed at which a retractable-gear aircraft may safely be flown with the gear in the extended position.

Landing-gear-operating speed—The maximum speed at which a retractable-gear aircraft may be flown while the gear is in the process of being extended (or retracted). On some aircraft it is lower than the landing-gear-extended speed because gear doors and related apparatus may be susceptible to damage from high-speed airflow when in the process of opening or closing.

Landing minimums—The minimums in terms of ceiling and visibility applicable to the published instrument approaches for any given airport.

Landing pattern—The standard pattern, consisting of downwind leg, base leg and final approach, which should be flown prior to and during the landing unless clearance for a straight-in or other approach has been given by the control tower.

Landing roll—The distance from touchdown to where the aircraft has been brought to a complete stop. Also called ground run or landing run.

Landing run—Landing roll.

Landing sequence—The order in which aircraft are positioned and follow one another for landing.

Landing site—In ballooning, any area free of obstructions which lends itself to the landing of a balloon.

Landing speed—The minimum speed at which an aircraft can touch down while under control by the pilot.

Lapse rate—The rate at which atmospheric temperature decreases with altitude. Under standard conditions this is 3.5 degrees per 1,000 feet. See adiabatic rate.

Large aircraft—An aircraft of more than 12,500 pounds maximum certificated takeoff weight, regardless of its actual operating weight at any given time.

Last assigned altitude—The last altitude or flight level assigned by ATC and acknowledged by the pilot.

LAT—Latitude.

Lateral axis—The imaginary line running from wingtip to wingtip through the center of gravity of the aircraft. Also known as pitch axis.

Lateral separation—The lateral spacing of aircraft flying at the same altitude, regardless of the direction of flight.

Latitude—Any line circling the earth parallel to the equator, measured in degrees, minutes and seconds north and south of the equator.

Piper Lance II (courtesy of Piper).

Launch site.

Latitude—Permissible deviation from the norm, such as, a pilot has a certain amount of latitude in selecting an rpm setting relative to the manifold pressure setting being used.

Launch site—In ballooning a large fairly flat area suitable for inflating the balloon and free of downwind obstructions.

Lawyer Pilots Bar Association—See *aviation organizations*.

Lazy eight—A coordination maneuver in which the nose of the aircraft is made to describe a pattern resembling a figure eight lying on its side on the horizon.

lb—Pound.

lbs—Pounds.

LC—Local telephone number of an FSS.

LCL—Local.

LCLZR—Localizer.

LCTD—Located.

LCTN—Location.

L/D—Lift-drag ratio; or lift over drag.

LDA—Localizer type directional aid.

Leading edge—The forward edge of any airfoil.

Lead/lag—See dragging.

Lead sled—In sailplane parlance a heavy glider with a poor glide ratio.

Lean—To adjust the fuel mixture so that it contains less fuel and more air.

LearAvia—Manufacturers of the *LearFan* turboprop aircraft. (Reno, NV 89506.)

LearFan—A revolutionary turboprop aircraft propelled by two turboprop engines driving a single pusher propeller mounted in the tail of the aircraft. The last design of the late William P. Lear. Developed and manufactured by LearAvia.

LearAvia turboprop aircraft specifications.

LEARAVIA CORPORATION MODEL:	LEARFAN 2100
ENGINES	PW/PT6B-35F (2)
shp (each)	650 flat rated
TBO (hours)	NA
SEATS crew + passengers	1+ 7 (max +0)
TAKEOFF ground roll (ft)	NA
50' obstacle (ft)	1,900
RATE OF CLIMB (fpm)	4,350 (1 eng. 2,210)
SERVICE CEILING (ft)	41,000 (1 eng. 33,000)
MAX SPEED (knots)	360 (19,000 ft)
BEST CRUISE SPEED (knots)	358 (25,000 ft)
LONG RANGE CRUISE SPEED (knots)	326 (39,000 ft)
RANGE w. res. best speed (nm)	1,630
long-range speed (nm)	2,010
STALL landing configuration (knots)	77
LANDING 50' obstacle (ft)	2,000
RAMP WEIGHT (lbs)	7,250
TAKEOFF/LANDING WEIGHT (lbs)	7,200/6,850
ZERO FUEL WEIGHT (lbs)	5,900
USEFUL LOAD (lbs)	3,250
FUEL CAPACITY (lbs)	1,700
MAX PAYLOAD (lbs)	1,900
PAYLOAD w. full fuel (lbs)	1,550
WING AREA (sq ft)	162.9
LENGTH/HEIGHT/SPAN external (ft)	39.7/NA/39.4
LENGTH/HEIGHT/WIDTH cabin (ft)	12.83/4.67/4.83
PRESSURIZATION (psi)	8.3
PRICE (1979 $s)	NA
STANDARD EQUIPMENT: VHF NAV and COM w. OBI; MARKER; GLIDE SLOPE; AUDIO PANEL; TRANSPONDER; ENCODING ALTIMETER; ADF; DME; RNAV; VNAV; AUTOPILOT; FLIGHT DIRECTOR; HSI; WEATHER RADAR; RADAR ALTIMETER; ANGLE OF ATTACK INDICATOR, GALLEY, TOILET.	

Lear Jet—The original spelling of the corporate jet aircraft designed and built by William P. Lear.

Learjet—A family of corporate jet aircraft manufactured by Gates Learjet Corporation.

Lear Siegler Instrument Company—Manufacturers of aircraft antennas and a variety of aircraft instruments. (4141 Easter Avenue SE, Grand Rapids, MI 49508.)

Learjet Longhorns during test flight (courtesy of Gates Learjet).

Lear Siegler Power Equipment Division—Manufacturers of alternators, generators and related equipment. (17600 Broadway, Santa Monica, CA 90406.)

Left-right indicator—On-course indicator.

Left seat—Flying the left seat is frequently used jargon for being pilot in command, regardless whether the crew consists of one or more persons.

Leg—A segment of a flight as in San Diego-Los Angeles leg.

Leg—Any of the four courses of an L/MF radio range.

Leg—Part of the traffic pattern; downwind leg, base leg.

Leigh Systems—Manufacturers of ELTs and related equipment. (6081 Court Street, Syracuse, N.Y. 13206.)

Lenticular cloud—A high cloud in the approximate shape of a double-convex lens, usually indicating severe turbulence below. Generally associated with mountain waves.

Letter of agreement—A written agreement between the FAA and the operator of an aircraft, permitting him to operate in a manner inconsistent with the regulations covering a given area of airspace.

Level action valve—In ballooning, a blast valve.

Level flight—Flight in which level altitude is maintained for a prolonged period of time.

Level turn—A turn without losing or gaining altitude, usually while maintaining a constant rate of turn and airspeed until the desired heading has been reached.

LF—Low frequency.

LF/MF radio range—A (largely antiquated) navigation facility which radiates four overlapping signal patterns. The overlapping areas are called legs, and the area between them, quadrants, identified by Morse code signals A (–) or N (–.).

LFR—LF/MF radio range.

LGT—Light.

LGTD—Lighted.

LGTG—Lightning.

Lift—The generally upward force created by the difference in pressure between the upper and lower surfaces of an airfoil in motion. In level flight it is balanced by the force of gravity. The force always acts at right angles to the flight path.

Lift component—The generally upward force created by the difference in pressure between the upper and lower surfaces of an airfoil in motion. In level flight it is balanced by the force of gravity. The force always acts at right angles to the flight path.

Lift component—The portion of lift which acts on the wing perpendicular to the direction of its motion through the air.

Lift drag ratio—(L/D)—The ratio of lift to drag of any structure. It is an indication of the efficiency of an airfoil at any given angle of attack.

Liftoff—The moment when the wheels break contact with the ground during takeoff.

Light aircraft—An aircraft with a maximum certificated gross weight of 12,500 pounds or less, regardless of the operating weight at any given moment.

Lighted airways—Airways, usually in remote parts of the country, marked by lights flashing course identification in Morse code. Remnants of the original airways system and, for all practical purposes, extinct.

Lighted airport—An airport where runway and obstruction lighting are available. There need not be a rotating beacon. At small airports such lighting may be available only on demand, either by contacting the nearest FSS (or unicom, if in operation), or in some cases by repeatedly depressing the mike button using a given transmitting frequency.

Lighter-than-air craft—Balloons, dirigibles, blimps.

Lighter-than-Air Society, Inc.—(LTA)—See *aviation organizations*.

Light gun—Located in control towers, it is a hand-held directional light-signaling device which emits a narrow bright signal of light (white, green, red). It can be operated by the tower controller in communicating with no-radio aircraft or aircraft experiencing radio failure. The color of the light and the frequency of flashes tells the pilot what action he is supposed to take. The pilot should acknowledge having received the message by rocking his wings or, when able, flashing his landing lights.

Light-light twin—An unofficial designation for twin aircraft with a maximum certificated takeoff weight of 4,000 pounds or less.

Lightning—Immensely powerful electrical discharges occurring in and near thunderstorms. Lightning can strike from cloud to ground or from cloud to cloud. Lightning strikes on aircraft are relatively rare but, when occurring, can cause damage to the aircraft skin and have been known to knock out electrical systems due to a momentary overload. Lightning usually occurs in areas of maximum turbulence and should be given a wide berth.

Light signals—The signals used by a tower controller using the light gun. Steady green means cleared for takeoff (aircraft on the ground) or landing (aircraft in the air); flashing green means cleared for taxi (aircraft on the ground) or return and land (aircraft in the air); steady red means stop (on the ground) or yield to other aircraft (in the air); flashing red means taxi clear of runway (on the ground) or do not land (in the air); flashing white means return to starting point (used only for aircraft on the ground).

Light twin—An unofficial designation for aircraft with certificated takeoff weights of between 4,000 and 6,000 pounds.

Lima—In aviation-radio phraseology the term used for the letter L. Pronounced leemah.

Limited remote communications outlet—An unmanned ground facility capable of forwarding transmissions and receptions to the associated VOR or VORTAC.

Limit-load factor—The maximum theoretical load factor which an airplane is expected to encounter. It is generally 3.8 G in a normal-category light plane and is well below the ultimate load factor, thus offering a substantial safety margin.

Line inspection—Preflight.

Line-of-sight—Literally the straight line from a person's eye to the horizon and from there out into space. Frequencies propagating according with the line-of-sight principle do not follow the curvature of the earth and are thus limited in reception distance. All VHF radio signals, both nav and com, fall into this category.

Line squall—A chain of cumulonimbus clouds along or ahead of a cold front, with violent winds, rain, lightning (and thunder) and the possibility of heavy hail. Such line squalls (lines of thunderstorms) may extend for hundreds of miles and tend to reach up very high. They cannot safely be overflown or circumvented except by certain turbine aircraft equipped with airborne weather radar. More often than not they don't last more than a few hours.

Liquid oxygen system—In these systems liquid oxygen is stored in supercooled (vacuum insulated) containers and can be turned into breathing oxygen (gas) when needed. The volumetric proportion between liquid and gaseous oxygen is 1 to 800.

LIRL—Low intensity runway edge lights.

Liter—A liquid or volumetric measure used in countries operating on the metric system. See conversion tables.

Litton Aero Products—Manufacturers of long-range navigation equipment. (21050 Burbank Boulevard, Woodland Hills, CA 91364.)

LKLY—Likely.

L/MF—LF/MF radio range.

LMM—Compass locator middle marker; part of an ILS.

LMT—Limit.

LND—Land.

LNDG—Landing.

Load—The total force acting on a structure, especially the wings and control surfaces of an airplane. It consists of static loads caused by gravity, dynamic loads resulting from certain maneuvers, or a combination thereof. See also payload, useful load, empty weight, gross weight, gust loads.

Load factor—The ratio of the total of loads exerted on an airplane to its original weight. A load factor of one G is in effect during level flight.

Load tapes—The horizontal and vertical tapes incorporated into the structure of a hot-air or gas balloon, which carry most of the load.

LOC—Localizer; part of an ILS.

Local aeronautical chart—A chart of a large terminal area at a scale of 1:250,000.

Local control frequency—Tower frequency.

Localizer—The portion of an ILS system which sends out directional VHF signals which show the pilot on his OBI and the action of the CDI whether he is aligned with the centerline of the runway, or off course to either left or right.

Localizer type directional aid—Similiar to a localizer, it is not associated with an ILS system, may not be used for precision approaches, and is frequently not aligned with the centerline of the runway.

Localizer usable distance—The maximum distance from which the localizer signal can be reliably received by an aircraft flying at a given altitude.

Local traffic—Aircraft operating in the traffic pattern and/or within sight of the tower, usually consisting of training flights and practice **VFR** or **IFR** approaches.

Locator beacon—ELT.

Lockheed Georgia Company—Manufacturers of the JetStar, the L-1011 and C-5A. (86 South Cobb Drive, Marietta, GA 33063.)

Log—A flight-by-flight record of all operations involving a specific aircraft, engine, or pilot, listing flight times, areas of operation, maintenance, repairs etc. There are generally three types of logs: The pilot logbook which must list all mandatory flight experience and flight tests, though logging other types of flying is voluntary. The airframe logbook which records the hours flown and all inspections, maintenance and repair work. And the engine log, which does the same for the engine.

Logbook—Log.

LOM—Compass locator outer marker; part of an ILS.

Lomcevak—A violent aerobatic maneuver in which the aircraft is made to tumble tail-over-prop through one and a-half revolutions. Only a handful of pilots have performed the true lomcevak, and not all aerobatic aircraft have the capability.

LONG—Longitude.

Longitude—Any line from the north to the south pole, measured in degrees, minutes and seconds of a circle, east or west of the Prime Meridian which runs roughly through Greenwich, England.

Longitudinal axis—An imaginary line running lengthwise through the airplane from the nose to the tail through the center of gravity. It is the axis about which the aircraft banks or rolls. Also called *roll axis*.

Longitudinal separation—The separation between aircraft flying at the same altitude and in (more or less) the same direction.

Long-range navigation—Navigation beyond the limits of the North American VOR/VORTAC system.

Long-range navigation systems—System capable of precision navigation independent of the VOR/VORTAC navigation network. They generally fall into two distinctly different types. One, known as INS (intertial navigation system) is controlled by a gyroscopic device and entirely independent of ground-based nav aids. The other, known as VLF, Omega or VLF/Omega receives position infcrmation from a limited number of very-low frequency radio transmitters, which information is

Long-range navigation systems comparision chart.

MANUFACTURER	MODEL	PRICE uninstalled	Current required	No. of waypoints	Weight lbs.	REMARKS
NORDEN	ONS VII	22,000	28V DC 115V AC 400 Hz	9	39	
CANADIAN MARCONI	CMA-734	25,000	28V DC	9	23.5	VLF $3,900 EXTRA
CANADIAN MARCONI	CMA-740	26,500	28V DC	9	34	VLF $3,600 EXTRA
LITTON	LTN-201	28,040	115V AC 400 Hz	9	34	
CANADIAN MARCONI	CMA-771	32,000	115V AC 400 Hz	9	40	
LITTON	LTN-211	32,500	115V AC 400 Hz	9	37	
BENDIX FLIGHT SYSTEMS	ONS-25	40,000	115V AC 400 Hz	9	30	COUPLES TO AUTOPILOT AND HSI
COMMUNICATIONS COMPONENTS CORPORATION	ONTRAC III	40,500	28V DC	10	27.2	
GLOBAL NAVIGATION, INC	GNS-500A	45,000	28V DC	10	38	
AIRESEARCH	AIRNAV 100	48,000	28V DC 115V AC 400 Hz	40	39.9	RHO/RHO (DME/DME) ALPHANUMERIC CDU
AIRESEARCH	AIRNAV 200	56,800	28V DC 115V AC 400 Hz	40	40.5	*2-WAY ELECTRIC DATA BUS INTERFACE WITH VLF/OMEGA/INS SYSTEM. PROVIDING NAV MANAGEMENT THROUGH SINGLE SYSTEM
LITTON	LTN-72	107,000	115V AC 400 Hz	9	62	
LITTON	LTN-72R	109,000	115V AC 400 Hz	9	62	
DELCO	C-IVA	112,000	115V AC 400 Hz	9	75.1	
TRACOR	7800	N.A.	28V DC 115V AC 400 Hz	9	38	TAS AVAILABLE BY TAS COMPUTER OR TAS PROGRAMMED INTO VLF OMEGA RPU

Inputs: altitude, VOR, DME, TAS, heading
Outputs: wind, HSI, ground sp., time to WP, error, track, position, INS, Omega, VLF, RNAV, non-volatile memory

then analyzed by a computer and the result displayed in the cockpit. Both systems can operate with extreme accuracy anywhere in the world.

LongRanger—A single-engine turbine helicopter of the JetRanger family, manufactured by Bell Helicopter Textron.

Loop—An aerobatic maneuver in which the airplane, without banking, describes a complete vertical circle in the sky. In an inside loop the head of the pilot points to the center of the circle. In an outside loop it points away from it.

Loop—A type of directional antenna used to receive low and medium frequency transmissions by the ADF.

Loop/sense antenna—A recent invention in the design of ADF antennas, eliminating the actual loop and reducing size and thus drag of the antenna installation.

Loran—Long range aid to navigation. A rather complicated low-frequency system involving fixes by triangulation between so-called master and slave stations. Rapidly becoming obsolete.

Lost communications—See two-way radio failure.

L over D—L/D

Low—A region of low atmospheric pressure surrounded by higher pressure. The winds around a low flow counterclockwise (in the northern hemisphere). Also called *cyclone* or *depression*.

Low altitude airway structure—The complete network of airways serving traffic up to, but not including, 18,000 feet msl.

Low altitude chart—Radio-facility chart for operations below 18,000 feet msl.

Low altitude en-route chart—Low altitude chart.

Low approach—An approach followed by a low flight over the runway, but without landing. Usually used after a practice instrument approach. Low approaches must have tower clearance at controlled airports.

Low clouds—Any clouds with bases below 6,000 feet agl.

Low frequency—Radio-magnetic waves with frequencies from 30 to 300 kHz (30,000 to 300,000 cycles per second).

Bell Long Ranger (courtesy of Bell).

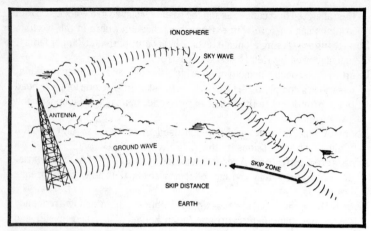

Low-frequency radio wave propagation (courtesy of FAA).

Low-frequency radio range—See LF/MF radio range.

Low-pressure system—A low.

Low tow—In sailplaning an airplane tow in which the sailplane flies below the flightpath of the tow plane (not recommended).

Low wing—Having the wing of a monoplane attached at or near the bottom of the fuselage.

LPG—Liquefied petroleum gas (propane).

LRCO—Limited remote communication outlet.

LRS—Lake reporting service. A special flight-watch service offered over the Great Lakes, the Long Island Sound and other areas where an emergency landing would be difficult if not impossible.

LSR—Light snow on the runway.

LTA—Lighter than air (craft).

LTA—Lighter-than-Air Society, Inc.

LTL—Little.

LTLCG—Little change.

LTR—Later.

Lubber line—The reference line on a magnetic compass or other instrument indicating heading. It is aligned with the longitudinal axis of the aircraft.

Lunar Rocket—A single-engine fixed-gear tailwheel STOL aircraft manufactured by Maule Aircraft.

LVL—Level.

LWR—Lower.

Lycoming—Avco Lycoming Division, manufacturers of reciprocating and turbine engines. (Avco Lycoming Engine Group, 652 Oliver Street, Williamsport, PA 17701.)

LYR—Layer.

M—Mach number.
m—Maritime air mass.
M—Measured (in sequence reports).
m—Meter(s).
M—Mike (phonetic alphabet).
M—Missing (in sequence reports).
MAA—Maximum authorized altitude.
Mach number—The ratio of TAS to the speed of sound. Mach 1 is the speed of sound under standard atmospheric conditions. Named after Ernst Mach. The following are some typical examples of the IAS/TAS in terms of subsonic Mach numbers at given altitudes under standard conditions:

Altitude	Mach. 76	.80	.84	.88	.92	.96
10,000	425/485	450/511	472/537	495/562	518/585	541/610
20,000	353/467	373/492	393/516	412/541	434/565	452/589
31,000	280/445	297/469	314/493	330/516	346/540	365/564
43,000	209/438	223/462	235/485	250/507	265/530	278/553

Mackerel sky—Cirrocumulus clouds in a large, rippling layer.
MacCready speed ring—A rotable bezel around a variometer, used in sailplanes to show the best speed to fly between thermals for the fastest speed in terms of lateral distance covered.
mag—Magneto.
MAG—Magnetic.
Magnetic compass—A compass which is always aligned with the magnetic north except during periods of turns, acceleration, deceleration, steep climbs.

Magnetic course—(MC)—True course corrected for magnetic variations.

Magnetic deviation—Error in the reading of the magnetic compass due to installation error or magnetic interference caused by cockpit instruments. It should be displayed in the cockpit on the compass correction chart.

Magnetic heading—True heading corrected for magnetic variations.

Magnetic north—The region, some distance from the geographic north pole where the earth's magnetic lines concentrate. A magnetic compass points to the magnetic north.

Magnetic variation—The angle between the magnetic and true north. It differs at various points on the earth due to local magnetic disturbances. It is shown on charts as isogonic lines marked with degrees of variation, either east or west, degrees which must be subtracted from or added to the true course to get the magnetic course. Easterly variations are deducted, westerly variations added (east is least, west is best).

Magneto—A self-contained generator which supplies electrical current to the spark plugs in the ignition system.

Main gear—The principal portion of the landing gear, usually two wheels located on either side of the fuselage, in tricycle-gear aircraft behind the center of gravity, and in tailwheel aircraft ahead of the center of gravity.

MAINT—Maintain.

MAINT—Maintenance.

Maintain—Concerning altitude, it is usually used by ATC in phrases like "Climb to and maintain..." or "Descend to and maintain. . ." At other times it might be used in such phrases as: "Maintain VFR."

Maintenance—Inspection, overhaul, repair, preservation, and replacement of parts. Different classes of aircraft and aircraft used either commercially or non-commercially are subject to different maintenance rules: but all aircraft must have one annual inspection and commercially used light aircraft must be inspected every 100 hours.

Main wheels—See main gear.

Make short approach—An ATC instruction, asking the pilot to enter final approach and land as quickly as possible, usually because of other traffic approaching to land. It is not a command. If the pilot feels that a short approach would be less than safe for him, he can refuse, in which case he may be asked to extend his downwind leg in order to permit the other traffic to land ahead of him.

MALS—Medium intensity approach light system.

MALSF—Medium intensity approach light system with sequenced flashers.

MALSR—Medium intensity lighting of simplified short approach light system with RAIL.

Maneuver—Any intentional change in the attitude or movement of an aircraft.

Maneuvering speed—The maximum speed at which abrupt control changes may be made without exceeding the load limits of the aircraft design.

Maneuvering vent—The vent near the apex of a hot-air or gas balloon, which can be opened and closed by the pilot in flight in order to increase or decrease lift.

Manifold pressure—The pressure of the fuel-air mixture in the intake manifold.

Manifold pressure gauge—The cockpit instrument showing the manifold pressure. It works on the principle of an aneroid barometer and reads in terms of inches of mercury (in Hg).

Manufacturers—Major manufacturers of aircraft, components, instruments and avionics are listed alphabetically along with their addresses as of 1979. A complete listing of all companies in any way involved with any phase of aviation may be found in the World Aviation Directory (Ziff-Davis Publishing Company, 1156 15th Street NW, Washington D.C. 20005. Published twice a year, spring and fall; single copy price is $40.00).

MAP—Missed approach point.

Marconi Elliott Avionics Systems—Manufacturers of a sophisticated line of avionics. (Christopher Martin Road, Basildon, Essex SS14 3EL, England.)

Maritime air mass—An air mass that originated over a large body of water and has a high moisture content concentrated in the lower layers. It may be polar or tropical.

Marker beacon—A 75 MHz transmitter which sends an elliptical signal pattern straight up with the size of the pattern expanding with altitude. It provides precise fixes at low altitudes.

Marker beacon receiver—A cockpit instrument which is designed to receive signals from marker beacons and to trigger visual and/or audio signals in the cockpit.

Marquise—One of the family of MU-2 corporate turboprop aircraft manufactured and marketed by Mitsubishi Aircraft International.

Master switch—A switch on the instrument panel of most aircraft, designed to turn the entire electrical system either on or off.

Marker beacon receiver (courtesy of King Radio).

Mitsubishi Marquise (courtesy of Mitsubishi).

Maule Aircraft—Manufacturers of single-engine high-wing fixed-gear STOL aircraft. (Spence Air Base, Moultrie, GA 31768.)

MAX—Maximum.

Max—Frequently used expression in combinations such as *max fuel, max payload,* etc.

Maximum allowable gross weight—The gross weight a particular aircraft is certificated to carry.

Maximum continuous horse power—Maximum cruise power.

Maximum flap extended speed—The maximum allowable structural speed at which an aircraft may be operated with flaps extended. It is shown as the upper limit of the white arc on the airspeed indicator.

Maximum fuel with a given payload—The maximum amount of fuel which may be carried in the tanks relative to the weight of passengers and/or freight being carried. Many aircraft are not designed to carry max fuel and max payload at the same time without exceeding maximum allowable gross weight.

Maximum payload—A greatest load in terms of weight, not counting the fuel and, in aircraft flown by a professional crew, the crew, which the aircraft can legally carry.

Maximum power—The greatest power available in an aircraft for use in emergencies.

Maximum structural cruise speed—The highest airspeed at which a given aircraft may be flown in rough air. Designated on the airspeed indicator by the upper limit of the green arc.

Mayday—An emergency voice message indicating to anyone receiving it anywhere in the world that the pilot is in trouble. Derived from the French phrase *m'aider*, meaning, "help me!"

MB—Marker beacon (receiver).

MB—Millibars.

mc—Megacycles. Now called MHz, megaHertz.

MC—Magnetic course.

MCA—Minimum cross altitude.

MCA—Minimum crossing altitude.

MCHP—Maximum continuous horsepower.

MDA—Minimum descent altitude.

MEA—Minimum enroute altitude.

Mean sea level—The sea level midway between the mean high and the mean low water, used as the standard reference point from which to measure altitude or elevation.

Measurements—See conversion charts.

Mechanically aspirated engine—Turbocharged or supercharged engine.

Medical certificate—The official proof or evidence that a pilot meets the medical standards required by the FARs. Examinations must be conducted by an approved aviation medical examiner. There are three classes of such certificates: A private pilot needs a third class certificate which must be renewed every 24 months. A commercial pilot needs a second class certificate, renewable every 12 months. Air transport rated pilots need a first class certificate, renewable every six months.

Medical examiner—A medical doctor authorized by the FAA to conduct medical examinations of pilots. Names and addresses can be obtained from the nearest FAA GADO office.

Medium frequency—The band of radio-magnetic waves propagating at frequencies of from 300 to 3,000 kHz (300,000 to 3,000,000 cycles per second.)

MEL(S)—Multi-engine land (sea); a pilot rating.

Melmoth—An all metal high-performance single-engine two-seat aircraft, designed and built by Peter Garrison. It became famous when he flew it first to Europe and then to Japan and back. It has since been turbocharged.

Mentor Radio—Manufacturers of avionics. (1561 Lost Nation Road, Willoughby, OH 44094.)

Mercator chart—A chart using the Mercator projection.

Mercator projection—A chart or map on which the meridians are drawn parallel to each other and the parallels of latitudes are straight lines whose distance from one another increases with the distance from the equator. Named after Gerhardus Mercator.

Meridian—Any line of longitude.

Merlin—A family of corporate turboprop aircraft manufactured by Swearingen Aviation Corporation.

A homebuilt Melmoth.

MES—Multi-engine sea; a pilot rating.

Messerschmitt-Boelkow-Blohm GmbH—Manufacturers of turbine helicopters. (Box 801109, 8000 Munchen 80, West Germany. U.S. representative: Boeing Vertol Company, P.O.Box 16858, Philadelphia, PA 19142.)

Meteorology—Science dealing with the atmosphere and its phenomena; specifically with weather and weather forecasting.

Meter—Metric means of measuring distance. See conversion charts.

Meter valve—In ballooning, part of the plumbing which meters the propane gas to the burner which heats the air in a hot-air balloon.

Metro—A turboprop commuter airliner manufactured by Swearingen Aviation Corporation.

MF—Medium frequency.

MFOB—Minimum fuel on board.

MH—Magnetic heading.

MH—Non-directional homing beacon operating on less than 50 watts.

MHA—Minimum holding altitude.

mHz—MicroHertz.

MHz—MegaHertz (1,000,000 cycles per second).

mi—Mile.

Micro Line—A family of high-quality avionics designed especially for high-performance singles and light twins by Collins Radio.

Microphones—The portion of a transmitting system which accepts the human voice (and other sound) and, via a diaphram or similar device, transforms it into transmittable electrical energy.

Homebuilt Melmoth specifications.

PETER GARRISON MODEL:	MELMOTH (homebuilt)
ENGINE (hp)	C/TSIO-360 (210)
TBO	1,400
PROPELLER	const. speed
number of blades	2
LANDING GEAR	retractable
PERFORMANCE FIGURES based on 2,100 lbs takeoff weight	
TAKEOFF ground roll (ft)	1,200 (est.)
50' obstacle (ft)	1,800 (est.)
RATE OF CLIMB (fpm)	1,800
Vx (knots)	95
Vy (knots)	105
SERVICE CEILING (ft)	30,000 (est.)
MAX SPEED (knots)	222 (30,000 ft)
CRUISE SPEED (knots)	75%:205 (24,000 ft)
economy (knots)	55%:170 (24,000 ft)
RANGE w. res. (nm)	75%:2,300 (12,500 ft)
economy (nm)	55%:2,850 (12,500 ft)
FUEL FLOW (pph)	75%:64
economy (pph)	55%:47
STALL clean/dirty (knots)	80/65
LANDING ground roll (ft)	1,000 (est.)
50' obstacle (ft)	2,000 (est.)
MAX TAKEOFF/LANDING WEIGHT (lbs)	2,950
USEFUL LOAD (lbs)	1,450
FUEL CAPACITY (lbs)	925
MAX PAYLOAD (lbs)	525
PAYLOAD w. full fuel (lbs)	525
FUEL W. MAX PAYLOAD (lbs)	925
WING LOADING (lbs per sq ft)	31.7
WING AREA (sq ft)	93
LENGTH/HEIGHT/SPAN external (ft)	21.5/8.3/23.3
LENGTH/HEIGHT/WIDTH cabin (ft)	9/3.9/3.7
SEATS	2 or 3
TURBOCHARGER make/type	Rajay/manual wastegate
PRICE	not for sale
EQUIPMENT:	
IFR panel; VHF COM and NAV w. OBI; MARKER; GLIDE SLOPE; AUDIO PANEL; TRANSPONDER; ENCODING ALTIMETER; EGT; CHT; ADF; DME; HF TRANSCEIVER; ANGLE OF ATTACK INDICATOR; AUTOPILOT; FUEL TOTALIZER	

Swearingen Merlin IV A (courtesy of Swearingen).

Messerschmitt-Boelkow-Blohm Bo 105.

Messerschmitt-Boelkow-Blohm twin-turbine helicopter specifications.

MESSERSCHMITT-BOELKOW-BLOHM MODEL	Bo 105S
ENGINES	AL/TA-63-250-C20B (2)
shp each	420
TBO (hours)	2,500
SEATS	5
RATE OF CLIMB max sea level (fpm)	1,650 (1 eng. 180)
SERVICE CEILING (ft)	17,000 (1 eng. 11,500)
HIGE (ft)	8,900
HOGE (ft)	6,400
MAX SPEED sea level (knots)	132
NORMAL CRUISE SPEED (knots)	103
Vne (knots)	146
FUEL CAPACITY (lbs)	1,020
RANGE w. full fuel (nm)	350
GROSS WEIGHT (lbs)	5,070
EMPTY WEIGHT (lbs)	2,810
USEFUL LOAD (lbs)	2,260
PAYLOAD w. full fuel (lbs)	1,040
LENGTH/HEIGHT/WIDTH external (ft)	27.7/9.67/8.2
cabin (ft)	4.3/4/4.9
MAIN ROTOR number of blades	4
diameter (ft)	32
TAIL ROTOR number of blades	2
diameter (ft)	6.2
PRICE (1979 $s)	649,500

Swearingen Metro (courtesy of Swearingen).

Microwave landing system—(MLS)—A type of instrument landing system which uses microwaves. It is more efficient than the ILS in that it increases the traffic-acceptance capability of an airport.

Middle clouds—Clouds with bases between 6,000 and 20,000 feet; usually lower in the winter and higher in the summer.

Middle marker beacon—A marker beacon located approximately 3,500 feet from the threshold of the runway in an ILS, triggering aural and / or visual signals in the cockpit.

Midfield RVR—The runway-visual-range numbers obtained by equipment located at or near the half-way point of the runway.

MIDIZ—Mid-Canada identification zone.

Mike—In aviation-radio phraseology the term used for the letter M.

Mike—Microphone.

Mile-high club—A humorous name for a fictitious organization of pilots who claim to have made love while at least one mile agl.

Millibar—A unit of pressure equal to a force of 1,000 dynes per square centimeter or 1/1000 of a bar. It is also equal to about .03 in Hg.

Military operations area—A portion of airspace designed to separate military operations from IFR traffic. Non-participating IFR traffic may traverse such areas only if ATC can assure safe IFR separation.

MIN—Minimum.

min—Minute.

Minimum control speed—The lowest speed at which a twin-engine aircraft is controllable with the critical engine inoperative and the other at takeoff power.

Minimum cross altitude—The minimum altitude at which an aircraft may cross a fix under IFR conditions, usually based on obstructions.

Minimum crossing altitude—Minimum cross altitude.

Minimum descent altitude—(MDA)—The lowest altitude measured in feet msl, to which an aircraft may descend under IFR conditions when executing a non-precision instrument approach. If no visual contact has been established with the airport or clearly airport-related terrain when MDA is reached, the pilot must maintain that altitude until reaching the missed approach point and then must execute a missed approach.

Minimum enroute altitude—(MEA)—The altitude between radio fixes which assures acceptable navigational radio signal reception and meets the necessary obstruction clearance requirements.

Minimum fuel—Means that the aircraft has sufficient fuel on board to reach its destination, assuming little or no undue landing delays. It is not a declaration of an emergency.

Minimum holding altitude—The lowest altitude prescribed for a holding pattern, meeting obstruction clearance requirements and acceptable radio-signal reception.

Minimum IFR altitudes—Unless otherwise indicated on the appropriate aviation charts, the minimum IFR altitude in non-mountain areas is 1,000 feet above the highest obstacle within a horizontal distance of five statute miles from the intended course. In mountainous areas it is 2,000 feet instead.

Minimum obstacle clearance altitude—(MOCA)—The lowest possible altitude between fixes on or off airways which meets obstacle-clearance requirements for that route segment. At this altitude radio-signal reception is guaranteed for only 22 nm from a VOR.

Minimum reception altitude—(MRA)—The lowest altitude required to receive reliable signals from nav aids needed for navigation in the particular route segment.

Minimum safe altitude—The minimum altitudes specified for various operations in heavily populated, lightly populated and unpopulated areas in the FARs Part 91. Instrument approach charts show minimum safe altitudes and emergency safe altitudes which provide for 1,000 foot obstacle clearance within a given distance of a nav aid. They are for emergency use only and do not guarantee reliable reception of radio signals.

Minimums—Plural of minimum; also and technically more correctly expressed as *minima*.

Minimum sink speed—The indicated airspeed at which a sailplane loses altitude most slowly, usually a little above stalling speed and below the best glide speed.

Minimum vectoring altitude—The lowest altitude in terms of feet msl at which an aircraft will be radar vectored by a controller except as authorized in conjunction with radar approaches, departures or missed-approach procedures. Minimum vectoring altitudes, sometimes lower than the published MEAs, are depicted on charts available to controllers but not normally to pilots.

Minute—(')—1/60 degree of a circle.

MISG—Missing.

Missed approach—A mandatory maneuver which must be initiated whenever an aircraft on an instrument approach has reached the missed approach point at MDA or DH without making visual contact with the airport. The missed-approach precedures to be followed at individual airports are depicted and described on instrument approach charts.

Missed approach point—(MAP)—The point shown on each published instrument-approach chart at which a missed approach must be initiated if visual contact with the airport has not been established.

Missed approach segment—The portion of the published instrument-approach procedure which must be flown in accordance with those published directions after the decision has been made to miss the approach.

Mission—Term used to describe a flight from takeoff to landing. Most often used in conjunction with business and corporate aviation activities.

Mission profile—The various numbers (fuel consumption, rate of climb, speed, rate of descent etc.) associated with a specific mission.

Mist—Thin fog giving a visibility in excess of one kilometer (⅝ mile).

Mist—Commonly used term for drizzle.

Mitchell—See Edo-Aire/Mitchell.

Mitsubishi Aircraft International, Inc.—Manufacturers of the MU-2 family of turboprop aircraft. A subsidiary of Mitsubishi Heavy Industries of Japan. (P.O. Box 3848, San Angelo, TX 76901.)

Mixture—The mixture of fuel and air necessary for combustion in reciprocating engines.

Mixture control—A control knob in the cockpit permitting the pilot to adjust the mixture of fuel and air.

MKR—Marker beacon.

Mitsubishi MU-2P (courtesy of Mitsubishi).

MITSUBISHI AIRCRAFT INTERNATIONAL, INC. MODEL:	SOLITAIRE	MARQUISE
ENGINES	GA/TPE 331-10-501M	GA/TPE 331-10-501M
shp (each)	778 (flat rated)	727 (flat rated)
TBO (hours)	3,000	3,000
SEATS crew plus passengers	2 + 7 (+ 9 max)	2 + 6 (+ 7 max)
TAKEOFF ground roll (ft)	1,825	1,550
50' obstacle (ft)	2,170	1,800
RATE OF CLIMB (fpm)	2,675 (1 eng. 675)	2,850 (1 eng. 770)
Vxse (knots)	145	145
Vyse (knots)	152	150
Vmca (knots)	99	93
ACCELERATE-STOP DISTANCE (ft)	3,300	2,750
SERVICE CEILING (ft)	33,000 (1 eng. 18,200)	35,500 (1 eng. 20,300)
MAX SPEED (knots)	308 (16,000 ft)	321 (20,000 ft)
BEST CRUISE SPEED (knots)	295 (20,000 ft)	313 (20,000 ft)
LONG-RANGE CRUISE SPEED (knots)	274 (31,000 ft)	302 (31,000 ft)
RANGE w. res. best speed (nm)	1,100 (20,000 ft)	1,160 (20,000 ft)
long-range speed (nm)	1,395 (31,000 ft)	1,600 (31,000 ft)
FUEL FLOW per engine, best speed (pph)	592	610
long range (pph)	416	388
STALL clean/dirty (knots)	100/76.5	97/73
LANDING ground roll (ft)	1,128	960
50' obstacle (ft)	1,880	1,600
RAMP WEIGHT (lbs)	11,625	10,520
ZERO FUEL WEIGHT (lbs)	9,950	9,700
TAKEOFF/LANDING WEIGHT (lbs)	11,575/11,025	10,470/9,955
USEFUL LOAD (lbs)	3,975	3,510
FUEL CAPACITY (lbs)	2,700	2,700
MAX PAYLOAD (lbs)	2,300	2,945
PAYLOAD w. full fuel (lbs)	1,275	810
FUEL w. max payload (lbs)	1,675	820
WING LOADING (lbs per sq ft)	65	59
WING AREA (sq ft)	178	178
LENGTH/HEIGHT/SPAN external (ft)	39.4/13.7/39.22	33.25/12.95/39.22
LENGTH/HEIGHT/WIDTH cabin (ft)	21.5/4.26/4.95	13.4/4.26/4.95
PRESSURIZATION (psi)	6	6
PRICE (1979 $s)	1,185,000	980,000
STANDARD EQUIPMENT:		
VHF NAV and COM; OBI; AUDIO PANEL; MARKER; GLIDE SLOPE; TRANSPONDER; ENCODING ALTIMETER; ADF; DME; RMI; AUTOPILOT; FLIGHT DIRECTOR; HSI; WEATHER RADAR; ENGINE SYNC; DE-ICING wings, props; GALLEY; TOILET (Marquise only).		

Mitsubishi Aircraft International turboprop aircraft. Specifications.

MLS—Microwave landing system.

mm—Millimeter; 1/1000 meter.

MM—Middle marker.

MOA—Military operations area.

MOCA—Minimum obstruction clearance altitude.

Mod—Term used for an alteration or modification of a standard aircraft.

Mode—A letter (or number in the military) assigned to particular pulse spacing of radar signals. Mode A is the standard mode used in the air traffic control radar beacon system (ATCRBS). Mode C implies altitude-reporting capability (in conjunction with an encoding altimeter or altimeter-digitizer combination).

Modified Lambert conformal conic projection—The standard Lambert conformal conic projection cannot be used in depicting the polar regions. It is therefore not used above the 80-degree parallel. The modified version allows for this and produces a chart representation usable for navigation.

Moment—Torque.

Moment—A figure representing the rotating force times its perpendicular distance from the axis of rotation. It is used in calculating the cg range of an aircraft.

Mooney Turbo 231 and 201 (courtesy of Mooney).

Moment arm—The distance from the axis of rotation, affecting the moment.

Mooney Aircraft Corporation—Manufacturers of a family of high-performance single-engine aircraft. (P.O. Box 72, Kerrville, TX 78028.)

Mooney single-engine aircraft specifications.

MOONEY AIRCRAFT CORPORATION SINGLE-ENGINE HIGH PERFORMANCE AIRCRAFT			
MODEL	**RANGER**	**201**	**TURBO 231**
ENGINE	L/O-360-A1D	L/IO-360-A3B6D	C/TSIO-360-GB
TBO (hours)	2,000	1,600	1,400
PROPELLER	Const. speed	Const. speed	Const. speed
BLADES	2	2	2
GEAR	Retractable	Retractable	Retractable
SEATS	4	4	4
TAKEOFF ground roll (ft)	815	880	1,220
50' obstacle (ft)	1,395	1,559	2,060
RATE OF CLIMB (fpm)	800	1,030	1,080
Vy (knots)	87	88	96
SERVICE CEILING (ft)	16,500	18,800	24,000
MAX SPEED (knots)	147	175	201
CRUISE SPEED at 75% (knots)	143 at 10,000 ft	169 at 8,000 ft	183 at 18,000 ft
at 55% (knots)	123 at 12,500 ft	152 at 14,000 ft	163 at 24,000 ft
Range w. 45 min res; 75% (nm)	654 at 10,000 ft	835 at 8,000 ft	950 at 18,000 ft
55% (nm)	816 at 12,500 ft	929 at 14,000 ft	1,135 at 12,000 ft
FUEL FLOW at 75% (pph)	56.4	64.8	67.8
at 55% (pph)	42	51.6	49.8
STALL clean (knots)	58	63	63
dirty (knots)	49	55	57
LANDING ground roll (ft)	595	770	1,147
50' obstacle (ft)	1,550	1,610	2,280
RAMP WEIGHT (lbs)	2,575	2,740	2,900
ZERO FUEL WEIGHT (lbs)	NA	2,356	2,468
TAKEOFF/LANDING WEIGHT (lbs)	2,575	2,740	2,900
USEFUL LOAD (lbs)	1,050	1,100	1,100
MAX FUEL (lbs)	312	384	432
MAX PAYLOAD (lbs)	800	800	800
FUEL w. max payload	250	300	300
PAYLOAD w. full fuel	738	716	668
WINGLOADING at gross (lbs per sq ft)	15.4	16.4	16.6
WING AREA (sq ft)	167	167	174.8
LENGTH/HEIGHT/SPAN external (ft)	23.2/8.3/35	24.7/8.33/35	25.4/8.3/36.1
LENGTH/HEIGHT/WIDTH cabin (ft)	NA	NA	9.5/3.7/3.6
TURBOCHARGER make/type	--	--	Rajay/325E 10-1
PRICE (1979 $s)	35,325	46,725	51,975
STANDARD EQUIPMENT:			
EGT GAUGE	--	operational group	operational group
CHT GAUGE	yes	yes	yes

Morse code—The internationally accepted code of audible or visual dots and dashes representing all letters of the alphabet and digits from zero to nine:

A •—	M ——	Y —•——
B —•••	N —•	Z ——••
C —•—•	O ———	0 —————
D —••	P •——•	1 •————
E •	Q ——•—	2 ••———
F ••—•	R •—•	3 •••——
G ——•	S •••	4 ••••—
H ••••	T —	5 •••••
I ••	U ••—	6 —••••
J •———	V •••—	7 ——•••
K —•—	W •——	8 ———••
L •—••	X —••—	9 ————•

Motorglider—A glider or sailplane with either a detachable or permanently built-in engine which can be shut off in flight whenever the pilot wants to soar.

Mountain flying—Flying in mountainous areas which requires specialized knowledge of the effects of density altitude, air currents and weather phenomena indigenous to the mountains.

Mountain wave—Turbulent movement of strong air currents on the lee side of mountain ranges. Mountain waves are often associated with updrafts some distance from the mountain range and have been responsible for various altitude records set by sailplanes. Often associated with lenticular clouds capping so-called rotor currents, they tend to result in extremely rough rides for light aircraft.

Movement area—Runways, taxiways and other portions of the airport used for moving aircraft rather than parking or tiedown.

Moving target indicator—An electronic means of emphasizing returns from moving targets on a radar display, capable of minimizing the confusing effect of ground clutter.

mph—Statute miles per hour.

mps—Meters per second.

MRA—Minimum reception altitude.

MRKD—Marked.

MRKG—Marking.

MRTM—Maritime.

MSA—Minimum safe altitude.

msl—Mean sea level.

MST—Mountain standard time.

MTBF—Mean time between failures.

MTCA—Minimum terrain clearance altitude.

MTI—Moving target indicator.

Multi-engine aircraft—Any aircraft with more than one engine.

Mitsubishi MU-2N (courtesy of Mitsubishi).

Multi-engine rating—The primary purpose of the training associated with obtaining a multi-engine rating is to know how to continue to safely control the aircraft in the event of a failure of one engine (or several engines in the case of three-, four-, or more-engine aircraft).

MU-2—A family of corporate twin-turboprop aircraft manufactured by Mitsubishi Aircraft International.

MUN—Municipal.

Murphy's Law—If anything can go wrong, eventually it will.

Mush—To sink in a nose-high attitude in which the aircraft is not able to effectively respond to normal control inputs.

Mushing—Flying along in a nose-high attitude at low speed, always just at the edge of a stall.

MVA—Minimum vectoring altitude.

N—Night (instrument approach charts).

N—November (phonetic alphabet).

NA—Not available (used frequently in aviation publications to indicate data not furnished by the manufacturer of a product). Also, not applicable.

NAA—National Aeronautical Association.

NAAA—National Agricultural Aircraft Association.

NAAS—Naval Auxiliary Air Station.

Nacelle—A streamlined housing or compartment on an airplane, especially engine enclosures.

NAFEC—National Aviation Facilities Experimental Center; an activity of the FAA, located in Atlantic City, NJ.

NAFI—National Association of Flight Instructors.

Narco Avionics—Manufacturers of a wide variety of general aviation avionics. (Commerce Drive, Ft. Washington, PA 19034.)

NAS—National Airspace System.

NAS Stage A—Refers to the entire complex of hardware and software comprising the ATC enroute system, such as radars, computers, computer programs, communication apparatus etc.

NAS Stage A alert function—A computer-controlled function capable of alerting controllers to potential traffic conflicts before they become critical.

NASA—National Aviation and Space Administration.

NASAO—National Association of State Aviation Officials.

NATA—National Air Transportation Association.

National Aeronautical Association—(NAA)—See *aviation organizations*.

National Agricultural Aviation Association—(NAAA)—See *aviation organizations*.

228

Nacelle (courtesy of Piper).

National Airspace System—(NAS)—The national aviation complex of the U.S., consisting of all elements, aircraft, airmen, airports, airspace, nav aids, communication facilities, traffic control facilities and personnel, aeronautical charts and information, weather information, rules, regulations, procedures, technical information and the FAA as a whole.

National Air Transportation Association—(NATA)—See *aviation organizations*.

National Association of Flight Instructors—(NAFI)—See *aviation organizations*.

National Association of Priest Pilots—See *aviation organizations*.

National Association of State Aviation Officials—(NASAO)—See *aviation organizations*.

National Aviation and Space Administration—(NASA)—A government agency concerned with space exploration and sophisticated research and development in the areas of aviation, space and many related subjects.

National Business Aircraft Association—(NBAA)—See *aviation organizations*.

National Flight Data Center—(NFDC)—A Washington-based FAA facility operating a central aeronautical information collection and dissemination service for government, industry and the aviation community. It publishes the Flight Data Digest.

229

National Flight Data Digest—(NFDD)—A daily publication of the flight data collected by the Flight Data Center.

Nationality mark—The letter or letters or combinations of letters and digits used to identify the nationality or country or registry of an aircraft:

AN	Nicaragua	3C	Equatorial Guinea
AP	Pakistan	D	Germany
3A	Monaco	EC	Spain
5A	Lybia	EI	Ireland
B	China	EJ	Ireland
5B	Cyprus	EL	Liberia
CC	Chile	EP	Iran
CF	Canada	ET	Ethiopia
CN	Morocco	F	France
CP	Bolivia	G	United Kingdom
CR	Portugal	9G	Ghana
CS	Portugal	HA	Hungary
CU	Cuba	HB	Liechtenstein
CX	Uruguay	HB	Switzerland
HC	Ecuador	00	Belgium
HH	Haiti	0Y	Denmark
HI	Dominican Republic	60S	Somali
HK	Colombia	PH	Netherlands
HL	Republic of Korea	PI	Philippines
HP	Panama	PJ	Netherlands Antilles
HR	Honduras	PK	Indonesia
HS	Thailand	PP	Brazil
HZ	Saudi Arabia	PT	Brazil
5H	Tanzania	7P	Lesotho
9H	Malta	8P	Barbados
I	Italy	7QY	Malawi
JA	Japan	4R	Ceylon
9J	Zambia	5R	Madagascar
JY	Jordan	9Q	Leopoldville
9K	Kuwait	8R	Guyana
LN	Norway	SE	Sweden
LQ	Argentina	SP	Poland
LV	Argentina	ST	Sudan
LX	Luxembourg	SU	United Arab Republic
LZ	Bulgaria	SX	Greece
9L	Sierra Leone	TC	Turkey
9M	Malaysia	TF	Iceland
N	United States	TG	Guatemala
5N	Nigeria	TI	Costa Rica
9N	Nepal	TJ	Cameron
OB	Peru	TL	Central African Republic
OD	Lebanon	TN	Brazzaville
OE	Austria	TO	South Yemen
OH	Finland	TR	Gabon
OK	Czechoslovakia	TS	Tunisia
TT	Chad	XW	Laos
TU	Ivory Coast	XY	Burma
TY	Dahomey	XZ	Burma
TZ	Mali	3X	Guinea

5T Mauretania	4X Israel
7T Algeria	5X Uganda
5U Burundi	YA Afghanistan
VH Australia	YI Iraq
VT India	YK Syria
5V Togo	YR Romania
6V Senegal	YS El Salvador
9V Singapore	YU Yugoslavia
4W Yemen	YV Venezuela
5W Western Samoa	5Y Kenya
6W Senegal	6Y Jamaica
XA Mexico	9Y Trinidad and Tobago
XB Mexico	ZK New Zealand
XC Mexico	ZL New Zealand
9XR Rwanda	ZM New Zealand
XT Upper Volta	ZP Paraguay
XU Cambodia	ZS Union of South Africa
XV Vietnam	ZT Union of South Africa
	ZU Union of South Africa

National Oceanic and Atmospheric Administration—(NOAA)—The government agency concerned with weather reporting and forecasting. Also publishers of aeronautical charts.

National Pilots Association—(NPA)—See *aviation organizations*.

National search and rescue plan—An agreement among various agencies to coordinate search and rescue efforts of all types.

National Transportation Safety Board—(NTSB)—The agency of the government which investigates all transportation-related accidents, and makes recommendations, based on its findings, to minimize the chance of a repetition of such accidents.

NATL—National.

Nautical mile—(nm)—6,076 feet or 1.15 statute mile. It is equal to one minute of latitude.

nav—Navigation.

Nav aid or navaid—Navigation aid such as VOR, NDB, etc.

Navajo—A family of piston-powered cabin twins manufactured by Piper Aircraft Corporation.

Navcom—Radio systems combining the functions of navigation and communication. Some navcoms use the receiver portions for both functions (often referred to as one-and-a-half systems) while others combine two entirely separate units in one box (referred to as one-plus-one or two-systems).

Navigable airspace—All airspace at or above the minimum altitudes prescribed in the FARs, plus the lower airspace required for takeoff and landing.

Navigation—Any one of several methods of getting an aircraft from one predetermined point to another and means of establishing present position.

Navigational aid—Nav aid.

Navigation lights—The lights at the wing tips and tail of the aircraft. A red light is located at the left wing tip, a green light at the right wing tip and a

Piper Navajo (courtesy of Piper).

Navcom with OBI (courtesy of Genave).

white light at the tail. Also known as position lights. Navigation lights are mandatory for night flight.

Navigator—In multi-person crews the third crew member who handles navigation-related chores.

Nav instruments—The various cockpit instruments designed to receive and display signals from groundbased nav aids.

Navion—A high-performance single-engine aircraft. (No longer in production).

VHF navcom equipment comparison chart.

MANUFACTURER	MODEL	PRICE uninstalled	Volts input	Frequencies com range	spacing kHz	nav range	spacing kHz	electric Frequ. mechanical read-out	Frequ. storage	no. of channels	Automatic radial	Automatic squelch	Marker B. plug in	RMI output	Glide slope	Units	Weight lbs.	REMARKS
GENAVE	ALPHA/200B	995	14	118-127.9	100	108-117.9	100									1	5	AUTOPILOT OUTPUT INCLUDES K1 205 VOR INDICATOR.
KING	KX 145	1,210	14	118-135.95	25	108-117.95	50									1	3.8	*INCLUDES 200 COM FREQUENCIES (118-127.95) IN NAV HEAD.
EDO-AIRE/ FAIRFIELD	RT-553	1,325	14 28*	118-135.95	50	108-117.95	50									1	5.5	*28V ADAPTER $90
GENAVE	GA/1000	1,395	14 28*	118-135.975	25	108-117.95	50									1	5.3	*28 V ADAPTER $50 AUTOPILOT OUTPUT
EDO-AIRE/ FAIRFIELD	RT-553A	1,525	14 28*	118-135.975	25	108-117.95	50									1	5.5	*28V ADAPTER $90
EDO-AIRE/ FAIRFIELD	RT-563	1,795	14 28*	118-135.95	50	108-117.95	50									1	7.7	*28V ADAPTER $90
KING	KX 170B	1,885	14 28*	118-135.975	25	108-117.95	50									1	N.A.	* 28 V ADAPTER $142 KI203, 204, 208 OR 209 NEEDED
KING		1,990	14 28*	118-135.975	25	108-117.95	50									1	N.A.	*28 V ADAPTER $142 KI203. 204. 208 OR 209 NEEDED
EDO-AIRE/ FAIRFIELD	RT-563A	1,995	14 28*	118-135.975	25	108-117.95	50									1	7.8	*28V ADAPTER $90
ARC	300/ RT-385A	2,495	28	118-135.975	25	108-117.95	50					OPT				1	7.5	
BENDIX AVIONICS	CN 2013 A	2,895	14 28	118-135.975	25	108-117.95	50									2	8.5	SINGLE NAV COM INCL. GLIDE SLOPE & MARKER
BENDIX AVIONICS	CN 2012 A	3,740	14 28	118-135.975	25	108-117.95	50		1							2	9	SINGLE NAV COM INCL. GUIDE SLOPE, MARKER AND FREQUENCY PRESELECT
ARC	400/ RT-485	4,195	28	118-135.975	25	108-117.95	50		3							1	7.8	
BENDIX AVIONICS	CNA 2010	6,155	14 28	118-135.975	25	108-117.95	50		1							3	14.6	DUAL NAVCOM W. GLIDE SLOPE, MARKER, FREQUENCY PRESELECT AND AUDIO CONTROL PANEL

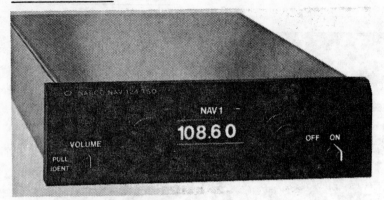

Nav receiver (courtesy of Narco).

Nav receiver—A radio receiver designed to receive signals from VORs.

NBAA—National Business Aircraft Association.

NDB—Non-directional beacon.

Negative—Expression used in aviation-radio phraseology in place of *no* or *not available*. For instance, *negative transponder* means that the aircraft is not equipped with a working transponder.

Negative contact—Phrase used by pilots to inform ATC that announced traffic is not in sight; or that the pilot has been unable to get a response on a given frequency.

Negative dihedral—The dihedral of the wing is downward from the lateral axis.

Negative lift—The force which will cause an airfoil to want to move downward rather than up. Many horizontal stabilizers are designed to produce a degree of negative lift.

Never-exceed speed—The highest speed at which an airplane may be flown in smooth air according to its certification parameters. Shown on the airspeed indicator by a red line. Also called *red-line speed*.

NFDC—National Flight Data Center.

NFDD—National Flight Data Digest.

NGT—Night.

Nicad battery—Nickel-cadmium battery.

Nickel-cadmium battery—A type of battery using the interaction of nickel and cadmium to generate electricity. It usually consists of a large number of individual cells, each producing between one and 1.5 volts. They are connected in sequence to produce the needed voltage. Nicad batteries are found primarily in turbine-powered aircraft. Charging procedures and maintenance are extremely critical. See *thermal runaway*.

Night—The time between the end of official evening twilight and the beginning of official morning twilight as published in the American Air Almanac.

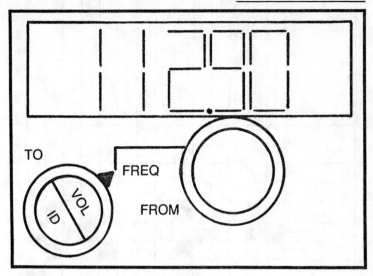

Nav receiver with digital frequency readout.

Night effect—The tendency of an ADF to fluctuate during dusk, night and dawn, when tuned to an appropriate nav aid or standard broadcast station.

Night error—Night effect.

Night flying—There is little difference between flying at night or during the day, assuming that the pilot is able to control the aircraft by means of

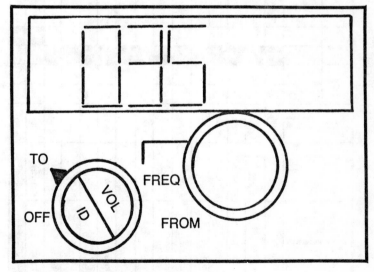

Nav receiver with digital omni-bearing readout.

VHF nav receivers comparison chart (continued on page 237).

(continued on page 237).

MANUFACTURER	MODEL	PRICE uninstalled	Volts input DC	AC 400 Hz	Mount panel	remote	frequency range MHz	Spacing kHz	Storage	VOR	GS	DME	RNAV output	Frequency readout electric	mechanical	Radial/bearing readout	RMI output	Units	Weight lbs.	REMARKS
MENTOR	M200	675	14/28*				108-117.95	50		200									2.7	*28V ADAPTER $96
MENTOR	M200/VL-2	910	14/28*				108-117.95	50		200						■		2	5.2	*28V ADAPTER $96 INCL. VOR/LOC CONVERTER-INDICATOR
COLLINS	VIR-350	1,065	14				108-117.95	50		200		■			■			1	2.5	BUILT-IN CONVERTER WILL DRIVE ANY STANDARD ARINC OBI OR HSI
NARCO	NAV 121	1,250	14/28				108-117.95	50		200	40							1	3	
EDO-AIRE/FAIRFIELD	R-552 / CID 552	1,270	14/28				108-117.95	50		200					■			2	5	
COLLINS	VIR-351	1,395	14				108-117.95	50		200					■	■		1	2.7	BUILT-IN CONVERTER WILL DRIVE ANY STANDARD ARINC OBI OR HSI
EDO-AIRE/FAIRFIELD	R-662 / CID 662	1,470	14/28				108-117.95	50		200					■			2	4.9	
RADAIR	200/240	1,595	14/28*				108-117.95	50		200	40				■			2	2.5	*28V ADAPTER NEEDED
NARCO	NAV 124	1,855	14/28				108-117.95	50		200	40				■			1	4.5	INCL. MARKER BEACON REC.
KING	KN 53	1,865	14/28				108-117.95	50	1	200	40			■				1	3.1	NON-VOLATILE MEMORY DIGITAL READOUT
EDO-AIRE/FAIRFIELD	R-554 / CID 554	1,935	14/28				108-117.95	50		200	40	■			■			2	5.4	
EDO-AIRE/FAIRFIELD	R-664 / CID 664	2,235	14/28				108-117.95	50		200	40	■			■			2	5.4	
NARCO	NAV 200	3,300				■	108-117.95	50	1	200	40			■				1	3.25	DUAL DISPLAY OF ACTIVE AND STORED FREQUENCIES DIGITAL TO/FROM BEARING INFO.
KING	KNR G32	3,465	28			■	108-117.95	50		200			■	■		■	■	1		
ELECTRONIQUE AEROSPATIALE	NR 810R	3,500	28				108-117.95	50		200	40				■			1	4.6	INCL. MARKER BEACON REC.

VHF nav receivers comparison chart.

Manufacturer	Model	Price	Voltage		Freq. Range			200	40	Qty	Weight	Comments
BECKER FLUGFUNKWERK	NR 2029	3,600	14/28		108-117.95	50		200	40	2		
	KNR 630	4,200	28	26	108-117.95	50		200	40	1	5.6	
BECKER FLUGFUNKWERK	NR 2030	4,300	14/28		108-117.95	50		200	40	2	4.8	
KING	KNR 615	4,465	28	26	108-117.95	50		200	40	1	7.3	
ARC	R-1048A	4,495	28		108-117.95	50	3	200		2	6.7	
ELECTRONIQUE AEROSPATIALE	NR-810	4,500	28		108-117.95	50	4	200	40	1	5.1	INCL. MARKER BEACON REC.
COLLINS	VIR-30M	5,160	28		108-117.95	50		200	40	1	5.7	INCL. MARKER BEACON REC. USE WITH ANY STANDARD OBI
KING	KNR 6030	5,410.50	28	26	108-117.95	50		200	40	1	11.9	
COLLINS	VIR-30A	5,545	28	26	108-117.95			200	40	1	5.8	INCL. MARKER BEACON REC. USE WITH ANY STANDARD OBI
KING	KNS-80	5,665	14/28		108-117.95	50		200	40	1	6	INCLUDES SELF-CONTAINED 4-WAYPOINT RNAV. DIGITAL READOUT
COLLINS	51RV-4	5,824	27.5	26	108-117.95	50		200	40	1	10.5	ARINC 547
BENDIX AVIONICS	RNA-34A	6,000	28.5	26	108-117.95	50		200	40	1	11.5	ARINC 547 VOR/ILS PLUS SERIAL TUNE INPUT AND DIGITAL BEARING OUTPUT
BECKER FLUGFUNKWERK	NR 2020/40	6,350	14/28		108-117.95	50		200	40	2	5.2	INCL. MARKER BEACON REC. DRIVES HSI
COLLINS	51 RV-4B	6,412	27.5	26	108-117.95	50		200	40	1	11.5	ARINC 547
ELECTRONIQUE AEROSPATIALE	DVR-740	6,600	115		108-117.95	50		200		1	11	ARINC 579
COLLINS	VIR-31A	6,820	28	26	108-117.95	50		200	40	1	7	ARINC AUTOMATIC NAV OUTPUT. HELICOPTER VERSION AVAILABLE
BENDIX AVIONICS	RVA-33A	7,064		115	108-117.95	50		200		1		NO MOVING PART ASSEMBLY ARINC 579 VOR
BENDIX AVIONICS	RIA-32A	8,648		115	108.1-111.95	50		78	40	1	14	ARINC 578 ILS
ELECTRONIQUE AEROSPATIALE	RNA-720	9,000	28		108-117.95	50		200	40	1	11	

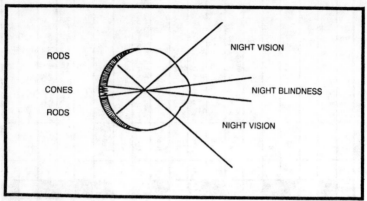

The human eye and night vision.

instruments alone, and that he can navigate with a minimum or total lack of outside visual cues. Night flight in single-engine aircraft adds the complication that in the event of an engine malfunction, an emergency landing cannot successfully be made in total darkness.

Night vision—The ability to use the rods rather than the cones of the eye by moving the direction of the eye from side to side to take advantage of the peripheral-vision capability of the rods and to minimize the effect of the blind spot straight ahead, caused by the inability of the centrally located cones to see under minimum-light conditions.

Nimbostratus—Middle stratiform cloud of a uniform dark grey color; usually associated with snow or steady rain.

Nimbus—A suffix or prefix (usually nimbo-) used with cloud types which are associated with rain.

Niner—In aviation-radio phraseology the term used for the number nine.

Ninety-Nines—An organization of women pilots originally founded by Amelia Earhart and 98 other female pilots. See *aviation organizations*.

nm—Nautical mile.

N-number—The identification number of an aircraft or U.S. registry. See also *nationality mark*.

NO—Number.

NOAA—National Oceanic and Atmospheric Administration.

No contact—Negative contact.

No-gyro approach—A radar approach provided pilots with inoperative or malfunctioning gyro instruments (DG, AH). In a no-gyro approach the controller will instruct the pilot: "Turn right (left)...stop turn."

No-gyro vector—A vector accomplished in the same manner as a no-gyro approach.

Noise—Sound at a level unpleasant to the human ear. The interior noise level of aircraft can eventually cause a degree of hearing loss. This can be avoided by wearing effective earplugs.

Noise—Visual clutter on a radar display, often caused by atmospheric disturbances.

Noise—The exterior noise created by certain aircraft during takeoff and/or landing is resulting at an increasing number of airports in attempts to institute some kind of curfew or to ban certain aircraft altogether from using the airport.

Noise regulation—FAR Part 36 spells out current and future regulations with reference to permissable exterior noise levels.

Noncontrolled airport—Uncontrolled airport.

Non-directional—Omni-directional; sending or receiving equally to or from all directions.

Non-directional beacon—(NDB)—A LF/MF nav aid sending non-directional signals which can be received by an ADF in the aircraft and used for navigation. Though not intended as such, standard broadcast stations can be used in the same way as NDBs.

Non-precision approach—Any instrument approach with no ground-based vertical guidance system such as a glide slope or precision approach radar. Non-precision approaches include localizer approaches (front and back course), VOR and ADF approaches, all circling approaches and a number of variations of the above.

Non-radar—Used as a prefix in connection with other terms such as: Non-radar approach, route, separation, etc., indicating that no radar coverage is available or being used.

Non-radar approach control—An approach-control facility without radar capability.

NOPT—No procedure turn is permitted; (on instrument approach charts).

NORAD—North American Air Defense Command.

Norden Systems—Manufacturers of long-range navigation equipment. (536 Broadhollow Road, Melville, NY 11746.)

NORDO—No-radio aircraft.

Normal climb—A climb at airspeed and power setting that produces the most efficient combination or gain in altitude and forward progress for the time and amount of fuel used.

Normal glide—A glide at an airspeed and angle that results in the greatest forward distance covered for a given loss of altitude. Also called *maximum glide*.

Normally aspirated—A non-turbocharded piston engine.

Normal operating range—The range of airspeeds within which a particular aircraft is supposed to be operated under normal conditions. It corresponds with the green arc on the airspeed indicator.

Normal operating range—With reference to instruments other than those associated with airspeed (tachometer, oil pressure, oil temperature, cylinder-head temperature, exhaust-gas temperature, etc.) the range within which it is safe to operate. Most such instruments also have a green marking of some kind to show the limits of the normal operating range.

Normal operating speed—Cruising speed.

Normal spin—A spin deliberately entered into by stalling the aircraft and from which recovery is made at the discretion of the pilot.

Normal stall—A deliberately created stall condition by slowly increasing the angle of attack, either with or without power.

North Atlantic Route—A coded route designed to utilize specific coastal fixes for trans-Atlantic flight.

Northerly turning error—The error in the reading of a magnetic compass when an aircraft in the northern hemisphere banks on a north-south heading. The vertical pull of the earth's magnetic field turns the compass card as the aircraft banks. An opposite effect takes place in the southern hemisphere.

Nose-heavy—Tendency of the aircraft to pitch forward when elevator control is released. Caused by faulty trimming or by forward-cg loading.

Nose-high—Flying with the nose of the aircraft higher than normal in relation to the level of flight. Usually caused by flying at a slow speed which requires a greater-than-normal angle of attack.

Nose-low—The exact opposite of nose-high; usually achieved at very high speeds, though barely noticeable.

Nosewheel—The (usually steerable) wheel under the nose of the fuselage in a tricycle-gear aircraft.

NOTAM—Notice to airmen.

Notice to airmen—An advisory distributed by the FAA giving current information about conditions or changes in any part or component of the National Airspace System.

November—In aviation-radio phraseology the term used for the letter N.

No voice—An indication on an aeronautical chart that a navigation facility has no capability for voice communication.

Nozzle—A narrow orifice through which fuel is fed to the carburetor or, in fuel-injection systems, into the combustion chamber.

NPA—National Pilots Association.

NSSFC—National Severe Storm Forecast Center in Kansas City.

NTSB—National Transportation Safety Board.

Null—The condition of a radio receiver when minimum reception is achieved.

Null position—The position of a loop antenna which produces minimum reception. In that position the sides of the loop antenna point to and from the station.

Numerous targets vicinity (fix)—An ATC traffic advisory to pilots, indicating a large number of radar returns and, in turn, a large number of aircraft at a given location.

NWS—National Weather Service.

O—Oscar (phonetic alphabet).

OAT—Outside air temperature.

OB—Olive Branch routes.

OBI—Omni-bearing indicator.

OBNR—Oil Burner route.

OBS—Omni-bearing selector.

Obscuration—A sky cover report indicating that the sky cannot be seen at all because it is obscured by surface phenomena such as fog, smog, smoke, rain, snow or dust extending upward from the ground. The sequence report symbol is X.

Obstacle—Buildings, chimneys, trees, powerlines, broadcast towers or other natural or man-made objects reaching to the height where they interfere with aviation.

Obstacle clearance altitude—Sufficient altitude to safely clear any obstacle in the vicinity.

OBSTN—Obstruction.

Obstruction—Obstacle.

Obstruction light—A light or group of lights, usually red, mounted on the surface of a structure or terrain feature to warn pilots of the presence of a flight hazard. In recent years many obstruction lights on broadcast towers and other high obstacles have been replaced with strobe lights which are easier to see under limited-visibility conditions.

OCAC—Oceanic air traffic control.

Occluded front—The front remaining on the surface after a cold front has overtaken a warm front. It may be cold or warm depending on which front displaces the other to the upper altitudes by the low-pressure motion at their point of intersection. It is indicated on weather charts by a line with alternating points and rounded marks. Also called *occlusion*, it can occasionally present a situation hazardous to light aircraft.

Occlusion—Occluded front.

OCT—Octane.

Octane rating—A measurement of the ability of fuel to burn smoothly and without detonation, compared to a pure test fuel called octane. Most aircraft require fuel with an octane rating in excess of 80. But since many companies have stopped producing 80-octane av gas, newer aircraft engines are designed to use low-lead 100-octane fuel.

OFAS—Overseas flight assistance service.

Off-route vector—An ATC radar vector which takes the aircraft along a route other than that previously assigned.

Offset parallel runways—Runways with parallel centerlines but with thresholds which are not aligned with one another.

Oil burner route—Areas with low-altitude high-speed military flight-training activity. The routes and hours of use are published in the special operations section of the Airman's Information Manual.

Oil canning—The rattling sound caused by the bulging in and out of fuselage skin during flight because of inadequate stiffeners. It is unpleasant but not necessarily dangerous.

Oil pressure gauge—An engine instrument which shows the pressure with which the lubricating oil is circulating through the moving parts of the engine. The readout is usually calibrated in psi, with the normal operating pressure indicated by a green bar or line.

Oil temperature gauge—An engine instrument which shows the temperature of the lubricating oil, indicating the normal operating range (usually by a green bar or line) and the point beyond which the temperature should not be permitted to rise. It can be used to judge engine temperatures in aircraft without cylinder-head-temperature gauge, though, when available, the latter is more accurate and to be preferred for that purpose.

Oleo strut—An assembly which dissipates shock by hydraulic action. Used primarily in landing gears to absorb the landing shock.

Olive branch routes—(OB)—Flight paths used by the Air Force and Navy in jet-aircraft training in VFR and IFR conditions, at altitudes from the surface to a published flight level. Information about the operational status of any particular route can be obtained by contacting the nearest FSS. Routes are described in AIM Part 4.

OM—Outer marker beacon.

Omega—A long-range navigation system in which the on-board navigation receiver uses the signals from a limited number of low-frequency transmitters to determine present position.

Omni—VOR.

Omni-bearing indicator—OBI)—The cockpit display used in VOR navigation, consisting of three components: 1)The omni-bearing selector

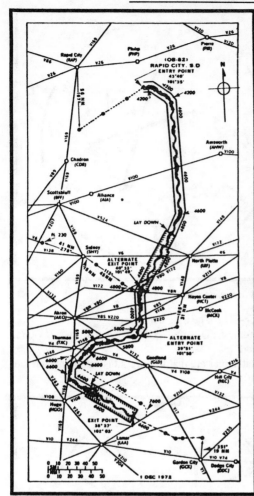

Typical oil burner route (courtesy of FAA).

(OBS) with which the pilot selects the radial from or bearing to a particular VOR. 2) The course-deviation indicator (CDI) consisting of a swinging needle or horizontally moving vertical bar which shows whether the aircraft is on the selected radial or bearing, or to either side of it. (Recently some manufacturers have begun to market CDIs with electronic displays. Typical of these is the Bendix 2000.) 3) The ambiguity or TO-FROM indicator which shows the pilot whether the selected radial is, in fact, a radial FROM the station or the bearing TO the station.

Omni-bearing selector—(OBS)—The element of the OBI used to select the bearing to or radial from the VOR

243

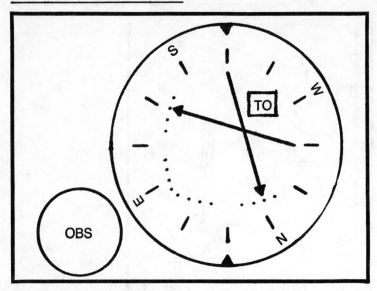

Traditional face of an OBI with glide slope.

Omni-course indicator—Omni-bearing indicator.

Omni range—The system of VORs and VORTACs on which the Victor airway system is based.

On course—Phrase used to indicate that the aircraft is established on the centerline of its route.

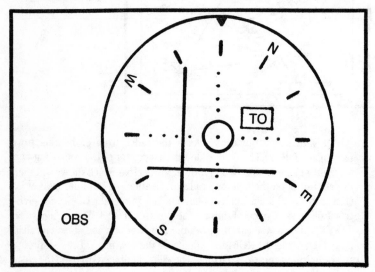

Newer version of the face of an OBI with glide slope.

Omni-bearing indicator (OBI). (courtesy of King Radio).

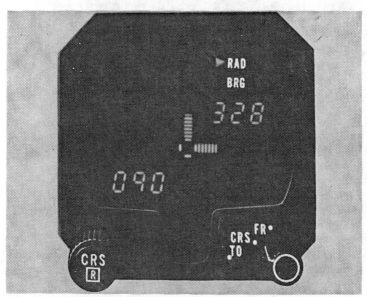

Bendix electronic course deviation indicator (courtesy of Bendix).

On course—Phrase used by ATC to inform a pilot making an instrument radar approach that he is correctly lined up with the final approach course.

On-course indicator—Any airborne instrument which shows the pilot that he is flying the desired course or what action he must take in order to intercept the desired course.

On instruments—Flying IFR.

On the gauges—Flying IFR.

Ontrac—Trade name for an omega long-range navigation system manufactured by Communications Components Corporation.

OPERG—Operating.

OPERN—Operation.

Optimum range—The longest distance a particular aircraft can travel, usually using higher altitudes and lower rpm settings than would be employed in normal cruise.

Option approach—An instrument approach, usually used for training purposes, which terminates either in a low approach, touch-and-go, missed approach or full-stop landing. See *cleared for the option*.

Organized Flying Adjusters—See *aviation organizations*.

Organized track system—A system of oceanic routes between Europe and Northern America which is adjusted twice daily to take advantage of the prevailing wind aloft.

Orientation flight—A flight made to permit a pilot to familiarize himself with a particular aircraft or a geographic area. Usually conducted in the company of a pilot familiar with either the aircraft or the area.

ORL—Overrun lights.

Oscar—In aviation-radio phraseology the term is used for the letter O.

Oscillation—A continuous movement of the aircraft around its pitch or lateral axis.

OSHA—Occupational Safety and Health Administration.

OSV—Ocean station vessel.

OTS—Out of service.

Out—In aviation-radio phraseology the term used to indicate that the conversation is ended and no further reply is expected.

OUTBND—Outbound.

Outbound—Flying away from the station or fix.

Outer compass locator—Compass locator co-located with the outer marker in an ILS system.

Outer fix—A general ATC term used with reference to terminal-area fixes other than those associated with the final approach.

Outer marker—Outer marker beacon.

Outer marker beacon—(OM)—A beacon, part of the ILS, located between four and seven miles from the threshold of the runway. It radiates a fan-shaped vertical signal and is identified in the cockpit by visual and/or aural signals of the marker-beacon receiver. It constitutes usually the beginning of the ILS final approach and aircraft should intercept it at an altitude which permits immediate interception of the glide slope.

Outside air temperature gauge—A thermometer mounted in such a way that it senses the temperature outside the fuselage and presents a readout inside the cockpit. Calibrated in either F. or C.

Outside loop—An aerobatic maneuver in which the aircraft executes a loop during which the pilot's head points away from the center of the loop. It may be entered into from normal flight by a push-over as for a steep dive, or from inverted flight with forward stick pressure throughout the maneuver.

Over—In aviation-radio phraseology the term used to imply that the transmission is ended but that a reply is expected.

Overcast—A sky-cover report indicating that the clouds cover more than 90 percent of the sky.

Overcontrol—To make large and usually erratic control movements which cause the aircraft to overreact. Also referred to as *chasing the needle*.

Overdevelopment—A term used primarily in soaring, indicating a rapid increase in the extent of cumulus cloud cover, reducing the sun's ability to heat the surface of the earth and thus slowing thermal activity.

Overhead approach—A series of maneuvers used by military pilots under VFR conditions as an entry into the traffic pattern.

Overload—To exceed the maximum allowable gross weight of a particular aircraft.

Overshoot—To fly beyond a designated or selected spot on the runway, frequently resulting in having the aircraft run off the end of the runway before it can be brought to a stop.

Overtaking—A faster aircraft approaching and then passing a slower aircraft from the rear. In this situation the aircraft being overtaken has the right of way and the overtaking aircraft should pass well to the right of the slower aircraft.

Over the top—VFR on top.

Owner-flown aircraft—In business and corporate aviation referring to aircraft which are usually flown by the owner or persons other than a professional crew.

Owner's handbook—Aircraft flight manual.

Owner's manual—Aircraft flight manual.

OXY—Oxygen.

Oxygen—A colorless, tasteless, odorless gas which forms about 21 percent of the atmosphere at sea level. The percentage of oxygen in the atmosphere decreases with altitude and supplemental oxygen is required when flying at higher altitudes. According to the FARs, supplemental oxygen must be used by the pilot (crew) when flying above 12,500 feet msl up to and including 14,000 feet msl for periods exceeding 30 minutes. When operating above 14,000 feet msl the pilot (crew) must use supplemental oxygen continuously; and when flying above 15,000 feet msl each occupant of the aircraft must be provided with supplemental oxygen.

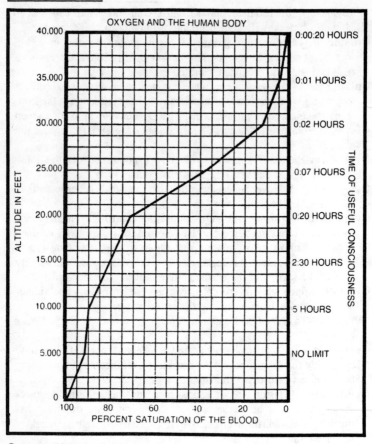

Oxygen and the human body.

oz—Ounce.

Ozone—A triatomic form of oxygen that is bluish, irritating gas of pungent odor. It is formed naturally in the upper atmosphere by the photochemical reaction with solar ultraviolet radiation.

Ozone layer—The atmospheric layer at an altitude between 20 and 30 miles msl, distinguished by a heavy concentration of ozone.

Ozonosphere—Ozone layer.

P—Papa (phonetic alphabet).

P—Polar air mass.

P—Precipitation ceiling (in sequence reports).

PA—Pressure altitude.

Pan—An emergency message indicating that the pilot needs help, but that the emergency is not as serious as it would be if he radioed *Mayday* rather than *Pan*.

Pancake landing—A landing in which the aircraft, usually unintentionally, is stalled somewhat too high above the landing surface and literally falls to the ground in a nose-high attitude. It can be used intentionally to reduce the landing roll on a short field, or in the case of an emergency off-airport landing, it will help to shorten the after-landing skid or roll on rough terrain.

Panel—Instrument panel.

Panel mounted—When used in relation to instruments or avionics, the term implies that the entire piece of equipment or most of its components are mounted behind the instrument panel. See also *remote mounted*.

PanTronics Corporation—Manufacturers of HF transceivers. (P.O. Box 22430, Ft. Lauderdale, FL 33335.)

Pants—Streamlined fairings enclosing the wheels of many fixed-gear aircraft, designed to reduce drag. Also called *spats*.

Papa—In aviation-radio phraseology the term used for the letter P.

PAR—Precision approach radar.

PAR approach—A precision instrument approach in which the controller, using PAR, gives instructions to the pilot with reference to the final approach course, the glide slope and the distance to the touchdown points. Today, few civil airports have PAR capability and it is used primarily by the military and in civil-aviation emergencies.

Pants, speed fairings, (courtesy of Piper).

Parachute—Originally, an umbrella-shaped fabric contraption of sufficient size to convey a person or freight safely to earth. In recent years parachutes have been designed in a variety of shapes, most intended to reduce the speed of descent and to increase the parachutist's ability to steer his chute to a predetermined landing spot.

Parallel—Latitude line.

Parallel entry—A suggested (but not mandatory) means of entering a holding pattern. The pilot approaches the holding fix from a direction more or less opposite of the direction in which the pattern is to be flown. He flies the first leg parallel to the holding pattern, makes a left procedure turn and then re-enters the holding pattern in the right direction.

Parallel offset course—A track which runs parallel to an established airway, either to the left or right, usually used by appropriately RNAV-equipped aircraft.

Parallel runways—Two or more runways at one and the same airport, the centerlines of which are parallel. At such airports the runways, in addition to the usual numbers, are identified with one of three letters: L for left, C for center, R for right. In some instances, such as at Los Angeles International Airport, where there are three or more parallel runways, one or two quite a distance from the other pair, the additional runways may be given a number one or two digits removed from the others, in order to avoid confusion. (At LAX the runways are 25L and 25R and 24L and 24R despite the fact that they are actually parallel.)

Parasite drag—All drag produced by aircraft surfaces which do not contribute to creating lift. It increases with an increase in airspeed.

Parking brake—A brake, usually a handbrake, which locks the wheels when the aircraft is parked or tied down.

Parrot—A slang term used for transponder.

Partial panel—A panel on which one or several of the basic flight instruments are either inoperative or covered up by the flight instructor. Executing certain maneuvers with a partial panel is part of instrument flight instruction.

Pathfinder 235—One of the Cherokee family of aircraft, single-engine, fixed gear, manufactured by Piper Aircraft Corporation.

PATN—Pattern.

Pattern—Landing pattern, holding pattern or any pre-determined or generally accepted manner in which to fly in order to achieve a given result.

PATWAS—Pilot's automatic telephone weather answering service.

Pawnee—A family of agricultural aircraft manufactured by Piper Aircraft Corporation.

Payload—Useful load minus the weight of fuel and oil and, in the case of professionally flown aircraft, the crew and its baggage.

PC—A term used by Mooney Aircraft Corporation to identify the permanently-on wing-leveler installation in its aircraft. It stands for *positive control*.

Piper Pawnee (courtesy of Piper).

PCA—Position control area.

PCPN—Precipitation.

PD—Period.

Peak envelope power—The maximum power in terms of watts, produced by a com transmitter.

PEP—Peak envelope power.

Performance chart—Charts, usually part of the aircraft flight manual, showing speeds, fuel consumption, rates of climb, minimum field lengths, etc. under a variety of engine-setting and other conditions.

Performance numbers—Octane ratings in excess of 100 in fuel designed for high-compression engines.

Performance parameters—The ranges in terms of speed, distance, service ceiling and so on, within which an aircraft is capable of performing safely.

Periodic inspection—The inspections of the airframe and engines and certain instruments which must be performed by authorized inspectors at intervals spelled out in the FARs.

Peripheral vision—The ability to see and recognize objects located to either side or above or below the direction in which we are looking. Especially important in night flight when the eye develops a blind spot straight ahead.

Permanent echo—A radar return, usually caused by fixed objects on the ground, representing a known location and sometimes used in directing aircraft to a desired position.

PERMLY—Permanently.

P-factor—An unevenness in thrust produced by the propeller during changes in aircraft attitude. It combines with torque to cause the aircraft to want to yaw in one direction or another during periods of maximum or minimum power. In typical U.S. aircraft in which the propeller turns to the right (as seen from the position of the pilot), the aircraft will try to yaw to the left (requiring right rudder) during full-power takeoff roll and climbout, and to the right (requiring left rudder) during a power-off glide to a landing.

Phonetic alphabet—The words used internationally for letters and digits:

A—Alpha	M—Mike	Y—Yankee
B—Bravo	N—November	Z—Zulu
C—Charlie	O—Oscar	0—Zero
D—Delta	P—Papa	1—Wun
E—Echo	Q—Quebec (Kebek)	2—Too
F—Foxtrot	R—Romeo	3—Tree
G—Gulf	S—Sierra	4—Fow-er
H—Hotel	T—Tango	5—Fife
I—India	U—Uniform	6—Six
J—Juliett	V—Victor	7—Sev-en
K—Kilo	W—Whiskey	8—Ait
L—Lima (Leemah)	X—Exray	9—Nin-er

Phugoid—The tendency of an aircraft to intermittently dive and climb, gaining speed during the dive portion until the added speed causes it to start to climb, losing speed until the cycle is repeated.

Piaggio—A family of Italian-built piston and turboprop aircraft, distinguished by a gullwing design and the fact that the propellers are aft mounted, operating as pushers.

PIC—Pilot in command.

Piccard Balloons—Manufacturers of hot-air balloons. (P.O. Box 1902, Newport Beach, CA 92663.)

Pictorial display—Any navigation-instrument display which attempts to pictorially represent the attitude of the aircraft with reference to the horizon or other fixed parameters. In contrast to displays using digits or needles to convey that information.

Pilatus Porter—A family of Swiss-built STOL aircraft.

Pilot—Airman. Anyone who flies a powered fixed-wing or rotary-wing aircraft, a glider, balloon, blimp or dirigible. (One who flies only hang gliders or uses high-performance or other types of parachutes is not a pilot.)

Pilotage—Navigation by reference to visible landmarks. Used usually in conjunction with aviation charts on which all meaningful landmarks are shown.

Pilot briefing—A service provided by flight service specialists at FSSs to help the pilot determine the weather and other conditions affecting a planned flight.

Pilot in command—The pilot responsible for the safe conduct of a flight. Regardless of the circumstances or of any ATC clearances or instructions, the pilot in command is responsible for the aircraft, passengers, crew and conduct of the flight.

Pilot light—On a hot-air-balloon heater the flame needed to activate the heater and replenish the supply of hot air and, in turn, lift.

Pilot-light valve—In a hot-air balloon, the valve which controls the pilot light.

Pilot report —(PIREP)—An observation of weather or other phenomena, their intensity, location and time made by a pilot in flight and reported to the nearest FSS which, at its discretion, may then transmit the report on the nationwide teletype system.

Pilot's automatic weather answering service—(PATWAS)—Prerecorded and periodically updated weather information, available at locations and during hours when on-site personnel is not available to brief pilots on flight conditions.

Pilot's discretion—An ATC phrase used with reference to altitude assignments. It means that the pilot may start his climb (or descent) whenever he wishes and that he may climb (or descend) at any rate of his choice. He may, if he so desires, temporarily level off at an intermediate altitude, but once having vacated a given altitude, he may not return to it.

Piper Chieftain and Navajo (courtesy of Piper).

Pilot self-briefing terminal—(PSBT)—A facility which gives pilots direct access to a computer for pre-flight briefing, obtaining information and filing flight plans. The pilot uses a keyboard telephone to convey his requests to the computer, and the computer replies either through alphanumeric displays on a TV screen or through computer-generated voice messages.

Pilot weather report—Pilot report.

Piper Aircraft Corporation—Manufacturers of a wide variety of single- and twin-engine piston and turboprop aircraft. (Lock Haven, PA 17725.)

PIREP—Pilot report.

Piston—Any structure which moves back and forth within a tube. More specifically, the circular ram which moves within the cylinder of reciprocating engines and represents the moving base of the combustion chamber. Also the moving inner part of an oleo strut.

Piston engine—Reciprocating engine.

Pitch—The attitude or movement of an aircraft with reference to or around its lateral axis; in other words, the position of the nose of the aircraft or its movement up or down. Also the blade angle of a propeller.

Pitch angle—Propeller-blade angle.

Pitch axis—Lateral axis.

Pitch-out—A sudden sharp turn away from the direction of flight, usually, but not necessarily, downward.

Pitot—A term associated with the airspeed-measuring systems. Named after Henry Pitot, a French physicist.

Pitot-heater—A means of heating the pitot-tube to avoid ice accumulation which would cause it to malfunction.

254

Piper Aircraft Corporation single-engine piston aircraft specifications.

PIPER AIRCRAFT CORPORATION MODEL:	TOMAHAWK	WARRIOR II	SUPER CUB	ARCHER II	DAKOTA	ARROW IV	TURBO ARROW IV	SIX 300	LANCE II	TURBO LANCE II
ENGINE	L/O-235-L2C	L/O320-D3G	L/O-320-D3G	L/O-360-A4M	L/O-540-J3A5D	L/IO-360-C1C6	C/TSIO-360-F	L/IO-540-K1G5D	L/IO-540-K1G5D	L/TIO-540-S1AD
TBO (hours)	NA	2,000	2,000	NA	2,000	1,600	1,400	2,000	2,000	1,800
PROPELLER	fixed pitch	fixed pitch	fixed pitch	fixed pitch	const. speed	const. speed	const. speed	const. speed	const. speed	const. speed
number of blades	2	2	2	2	2	2	2	2	2	2
LANDING GEAR	fixed, tricycle	fixed, tricycle	fixed, tail wheel	fixed, tricycle	fixed, tricycle	retractable	retractable	fixed, tricycle	retractable	retractable
SEATS	2	4	2	4	4	4	4	7	6	6
TAKEOFF ground roll (ft)	820	975	200	870	886	1,025	1,110	900	1,450	1,410
50' obstacle (ft)	1,460	1,490	500	1,625	1,216	1,600	1,620	1,350	2,360	1,875
RATE OF CLIMB (fpm)	718	710	960	735	1,110	831	940	1,050	1,000	1,050
SERVICE CEILING (ft)	12,000	14,000	19,000	15,000	17,500	17,000	20,000 (certif.)	17,100	15,400	20,000 (certif.)
MAX SPEED (knots)	109 (sea level)	126 (sea level)	113 (sea level)	133 (sea level)	148 (sea level)	152 (sea level)	178 (14,000 ft)	156 (sea level)	NA	183
CRUISE SPEED (knots)	75%:108 (7,100 ft)	75%:127 (9,000 ft)	75%:100	75%:131	75%:144	75%:143	75%:172 (14,000 ft)	75%:152	75%:158	75%:176
economy (knots)	65%:100 (10,500 ft)	65%:103 (12,500 ft)	55%:NA	55%:107	55%:130	55%:131	55%:154 (14,000 ft)	55%:132	55%:139	55%:142
RANGE w res. (nm)	75%:452 (7,000 ft)	75%:520 (9,000 ft)	75%:400	55%:515	55%:623	55%:810	55%:675 (14,000 ft)	75%:679 (8,000 ft)	55%:656	75%:597
economy (nm)	65%:468 (10,500 ft)	65%:635 (12,500 ft)	55%:NA	55%:670	55%:795	55%:875	55%:860 (14,000 ft)	55%:835 (12,000 ft)	55%:714	55%:623
FUEL FLOW (pph)	75%:NA	75%:60	75%:54	75%:63	75%:81.6	75%:63	75%:84	75%:108	75%:108	75%:131.4
economy (pph)	65%:NA	65%:54	55%:NA	55%:54	55%:55.8	55%:48	55%:55.2	55%:71.4	55%:71.4	55%:82.8
STALL clean/dirty (knots)	48/47	48/43.6	NA/37	55/37.8	65/56	60/55	63/56	62/55	53/52	57/52
LANDING ground roll (ft)	625	595	350	935	932	615	645	630	880	1,050
50' obstacle (ft)	1,465	1,115	885	1,400	1,410	1,525	1,620	1,000	1,710	1,760
RAMP WEIGHT, TAKEOFF/LANDING WEIGHT (lbs)	1,670	2,325	1,750	2,550	3,000	2,750	2,900	3,400	3,600	3,600
EMPTY WEIGHT/USEFUL LOAD (lbs)	1,088/582	1,344/981	946/804	1,412/1,338	1,633.5/1,366.5	1,593/1,157	1,638/1,262	1,837/1,563	1,968/1,632	2,065/1,535
FUEL CAPACITY (lbs)	180	288	214.8	288	432	432	462	588	564	564
MAX PAYLOAD (lbs)	NA	NA	NA	NA	880	880	880	1,420	1,220	1,220
PAYLOAD w full fuel (lbs)	NA	NA	NA	NA	808	695	800	975	922	714
FUEL w. max payload (lbs)	NA	NA	NA	NA	360	277	462	916	343	315
WING LOADING (lbs per sq ft)	13.39	13.7	10	15	17.6	16.2	17	19.5	20.6	20.6
WING AREA (sq ft)	124.7	170	178.5	170	170	170	170	174.5	174.5	174.5
LENGTH/HEIGHT/SPAN external (ft)	23.1/9.06/34	23.8/7.3/35	22.5/6.7/35.3	23.8/7.3/35	24/7.4/35	25/8/35	25/8/35	27.7/8.2/32.8	28.3/9.5/32.8	29/9.5/32.8
LENGTH/HEIGHT/WIDTH cabin (ft)	5.7/4.2/3.5	8.08/4.08/3.46	NA	8.08/4.08/3.46	8.08/4.08/3.46	7.9/4.1/3.46	7.9/4.1/3.46	10.4/4.1/4.1	NA	10.4/4.1/4.1
TURBOCHARGER	NA	NA	NA	NA	NA	NA	Yes	NA	NA	fixed waste gate
PRICE (1979 $s)	16,840	24,040	24,520	29,710	39,910	44,510	49,150	52,030	64,060	72,760

Piper Aircraft Corporation twin-engine piston aircraft specifications.

PIPER AIRCRAFT CORPORATION MODEL:	SEMINOLE	SENECA II	AZTEC F	TURBO AZTEC F	AEROSTAR 600A	AEROSTAR 601B	AEROSTAR 601P	NAVAJO	NAVAJO C/R	CHIEFTAIN
ENGINES	LIO-360-E1AD	C/TSIO-360-E	LIO-540-C4B5	LTIO-540-C1A	LIO-540-K1J5	LIO-540-S1A5	LIO-540-S1A5	LTIO-540-A2C	LTIO-540-F2BD (LT10)	JTIO-540-J2BD
TBO (hours)	2,000	1,400	NA	NA	2,000	1,800	1,800	1,800	1,800	1,800
PROPELLERS	const. speed ft	const. speed ft	const. speed ft	const. speed ft	const. speed ft	const. speed ft	const. speed ft	const. speed ft	const. speed ft	constant speed ft
number of blades, each	2	2	2	2	3	3	3	3	7	3
SEATS	4	6	6	6	6	6	6	7	7	8
TAKEOFF ground roll (ft)	880	900	990	990	1,008	1,000	1,000	1,030	990	1,360
50' obstacle (ft)	1,400	1,240	1,980	1,980	1,400	2,490	2,490	2,190	2,080	2,490
RATE OF CLIMB (fpm)	1,340 (1 eng. 217)	1,340 (1 eng. 225)	1,400 (1 eng. 235)	1,470 (1 eng. 225)	1,800 (1 eng. 450)	1,530 (1 eng. 254)	1,530 (1 eng. 254)	1,445 (1 eng. 245)	1,500 (1 eng. 255)	1,390 (1 eng. 230)
Vmc (knots)	69	69	70	70	80	84	80	74	74	78
Vsse (knots)	NA	82	84	84	104	96	100	97	97	107
Vyse (knots)	NA	90	91	91	113	100	109	96	100	108
SERVICE CEILING (ft)	17,100 (1 eng. 4,100)	25,000 (1 eng. 13,400)	17,600 (1 eng. 4,800)	24,000 (1 eng. 17,000)	21,000 (1 eng. 6,150)	25,000 (1 eng. 9,300)	30,000 (1 eng. 9,300)	26,300 (1 eng. 15,200)	26,400 (1 eng. 15,300)	27,200 (1 eng. 13,700)
CRUISE SPEED (knots)	75%: 166 (8,000 ft)	75%: 190 (20,000 ft)	75%: 179 (3,800 ft)	210 (22,000 ft)	75%: 220 (7,500 ft)	75%: 257 (25,000 ft)	75%: 257 (25,000 ft)	75%: 215 (22,000 ft)	75%: 220 (20,000 ft)	75%: 221 (20,000 ft)
economy (knots)	55%: 153 (16,000 ft)	55%: 165 (25,000 ft)	55%: 170 (6,250 ft)	193 (24,000 ft)	55%: 200 (10,000 ft)	55%: 209 (25,000 ft)	55%: 218 (25,000 ft)	55%: 186 (24,000 ft)	55%: 180 (16,000 ft)	55%: 177 (15,000 ft)
RANGE w. res. standard tanks (nm)	75%: 690 (7,000 ft)	55%: 546 (25,000 ft)	695 (2,000 ft)	695 (24,000 ft)	75%: 1,103 (10,000 ft)	65%: 1,086 (15,000 ft)	65%: 1,063 (15,000 ft)	55%: 1,005 (20,000 ft)	55%: 940 (20,000 ft)	55%: 885 (20,000 ft)
standard tanks (nm)	55%: 730 (14,500 ft)	55%: 609 (20,000 ft)	903 (7,000 ft)	835 (20,000 ft)	55%: 1,320 (10,000 ft)	55%: 1,271 (20,000 ft)	55%: 1,244 (15,000 ft)	55%: 1,065 (20,000 ft)	55%: 1,040 (16,000 ft)	55%: 950 (15,000 ft)
long-range tanks (nm)	NA	75%: 793 (20,000 ft)	1,060 (6,000 ft)	947 (22,000 ft)	NA	NA	NA	NA	NA	NA
long-range tanks (nm)	NA	55%: 882 (20,000 ft)	1,320 (7,000 ft)	1,145 (20,000 ft)	NA	NA	NA	NA	NA	NA
FUEL FLOW per engine (pph)	75%: 67.2	75%: 70.8	h speed 72	h speed 77	75%: NA	65%: 84	65%: 82	75%: NA	75%: NA	75%: NA
economy (pph)	55%: 44.4	55%: 54	economy 53	economy 63	55%: 68	55%: 71.5	55%: 72	55%: 72	55%: 72	55%: 72
STALL dependnty (knots)	57.55	63-61	61.54.5	61.54.5	70.74	NA.77	NA.77	NA	NA	NA
LANDING ground roll (ft)	385	920	760	760	892	1,730		906	906	1,045
50' obstacle (ft)	1,190	2,090	1,585	1,585	2,410	2,030	2,030	1,818	1,818	1,880
ACCELERATE-STOP DISTANCE (ft)	2,070	2,520	5,200	5,200	5,500	6,000	6,000	6,536	6,536	6,536
RAMP WEIGHT (lbs)	3,800	4,570	5,200-4,940	5,200-4,940	5,500	6,000	6,000	NA	NA	NA
TAKEOFF/LANDING WEIGHT (lbs)	3,800	4,570	1,979	1,842	1,720	2,015	1,997	NA	NA	NA
FUEL standard tanks (lbs)	660	588	823	823	1,047	1,047	1,027	1,123.8	1,123.8	1,123.8
long-range tanks (lbs)	NA	768	1,063	1,063	NA	NA	NA	NA	NA	NA
WING LOADING (lbs per sq ft)	21.11	22	25.1	25.1	32.4	33.7	33.7	28.4	28.4	30.6
WING AREA (sq ft)	180	208.7	207.6	207.6	170	178	178	229	229	229
LENGTH/HEIGHT/SPAN external (ft)	27.68.538.55	28.9/38.9	31.2.10.1.37.3	31.2.10.1.37.3	34.8.12.1.34.2	34.8.12.1.36.7	34.8.12.1.36.7	32.6/13/40.7	32.6/13/40.7	34.63/13/40.67
LENGTH/WIDTH/HEIGHT cabin (ft)	8.06/4.06/3.45	10.4/08-4.08	8.5.3.75.4.2	8.5.3.75.4.2	12.5.4.3.83	12.5.4.3.83	12.5.4.3.83	10.94.294.17	10.94.294.17	12.56/4.294.17
TURBOCHARGER	NA	fixed waste gate	NA	Yes	NA	Raply dual autom.	Raply dual automatic	Yes	Yes	Yes
PRESSURIZATION (psi)	NA	NA	NA	NA	NA	NA	4.25	NA	NA	NA
PRICE standard tanks (1979 $s)	80,260	99,640	127,120	147,925	178,700	257,800	205,200	208,540	222,060	239,320
long-range tanks (1979 $s)	NA	add 1,365	add 2,520	add 2,520	NA	NA				
STANDARD EQUIPMENT:										
VHF COM and NAV w. OBI, AUDIO PANEL	No	No	No	No	Yes	Yes	Yes	No	No	No
MARKER, GLIDE SLOPE	No	No	No	No	Yes	Yes	Yes	No	No	No
TRANSPONDER, ENCODING ALTIMETER	No	No	No	No	Yes	Yes	Yes	No	No	No
ADF, DME	No	No	*o	*o	No. Yes	Yes	Yes	*o	No	No

Piper Aircraft Corporation turboprop aircraft specifications.

PIPER AIRCRAFT CORPORATION MODEL:	CHEYENNE I	CHEYENNE II
ENGINES	PW/PT6A-11 (2)	PW/PT6A-28 (2)
shp each	500 flat rated	620 flat rated
SEATS crew + passengers	2 + 6	2 + 6
TAKEOFF ground roll (ft)	1,712	1,410
50' obstacle (ft)	2,541	1,980
RATE OF CLIMB (fpm)	1,750 (1 eng. 413)	2,800 (1 eng. 660)
Vxse (knots)	NA	104
Vyse (knots)	NA	117
Vmca (knots)	NA	96
ACCELERATE-STOP DISTANCE (ft)	3,300	3,300
SERVICE CEILING (ft)	28,200 (1 eng. 12,500)	31,600 (1 eng. 14,600)
MAX SPEED (knots)	249 (12,000 ft)	283 (11,000 ft)
BEST CRUISE SPEED (knots)	249 (12,000 ft)	283 (11,000 ft)
LONG-RANGE CRUISE SPEED (knots)	236 (25,000 ft)	250 (29,000 ft)
RANGE w. 45 min res., best speed (nm)	940 (12,000 ft)	900 (12,000 ft)
long-range speed (nm)	1,260 (12,000 ft)	1,510 (29,000 ft)
STALL SPEED clean/dirty (knots)	84/72	86/75
LANDING ground roll (ft)	1,695 (1,193 prop. rev)	1,430 (995 prop. rev)
50 ' obstacle (ft)	2,548 (2,131 prop. rev)	2,480 (1,860 prop. rev)
RAMP WEIGHT (lbs)	8,750	9,050
TAKEOFF/LANDING WEIGHT (lbs)	8,700/8,700	9,000/9,000
ZERO FUEL WEIGHT (lbs)	7,200	7,200
USEFUL LOAD (lbs)	3,850	4,074
FUEL CAPACITY (lbs)	2,010	2,559.4
MAX PAYLOAD (lbs)	1,330	1,255
PAYLOAD w. full fuel (lbs)	1,330	1,114.6
FUEL w. max payload (lbs)	2,010	2,449
WING LOADING (lbs per sq ft)	38	39.3
WING AREA (sq ft)	229	229
LENGTH/HEIGHT/SPAN external (ft)	34.67/12.75/40.67	34.67/12.75/42.68
LENGTH/HEIGHT/WIDTH cabin (ft)	8.42/4.29/4.17	8.42/4.29/4.17
PRESSURIZATION (psi)	5.5	5.5
PRICE (1979 $s)	540,280	635,890

Pitot-pressure—The pressure produced by an aircraft moving through the air. It increases with airspeed and is the basis for airspeed measurement. Also called *ram* or *impact air pressure.*

Pitot-static system—A device which compares pitot pressure with static or atmospheric pressure and presents the result in the cockpit terms of indicated airspeed, altitude and vertical speed. It consists of a pitot tube mounted on the frontal surface of the aircraft or wing and static vents placed flush into the wall of the fuselage. Both parts are needed to operate the airspeed indicator, and the static vents control the altimeter and the vertical-speed indicator.

Pitot tube—A tube exposed to the airstream, designed to measure the pressure with which an aircraft meets the static air. It must keep clear of obstructions and should be covered when the aircraft is not in use, as insects like to nest in it. Also called *pitot-static head.*

Pitts Aviation Enterprises—Manufacturers of highly successful aerobatic biplanes, used in national and international competitions, where they have been instrumental in obtaining many records. (P.O. Box 548, Homestead, FL 33030.)

Pitts Special—An aerobatic biplane.

Placard—Any sign or message, usually prominently displayed in the cockpit, to remind or warn the pilot to stay within certain performance perameters. An aircraft may be said to be *placarded* against intentional spins.

Planning chart—An aeronautical planning chart used prior to but normally not during a flight. A VFR planning chart, covering the conterminous 48 states, is available from the NOAA (or through the chart services of the AOPA or NPA). Also IFR (radio-facility) planning charts are available from the same sources or from Jeppesen-Sanderson, Inc.

PLD—Payload.

P-line—Pole line.

Plotter—A combination ruler and protractor, marking nm and sm scales, and used to determine or draw projected courses on charts.

Plumbing—In relation to aircraft, the term usually refers to the hydraulic system. In hot-air ballooning it refers to the propane tanks and the system of pipes and valves which deliver the propane to the burner.

PNR—Point of no return.

Pod—Nacelle.

Point of no return—The position along a line of flight, having passed beyond which the pilot cannot return to his point of departure on the amount of fuel remaining on board.

Polar air mass—A cold air mass, continental or maritime, that originated in the polar regions.

Polar easterlies—The general circulation pattern of easterly winds between 60 degrees N latitude and the north pole.

Polar front—A frontal zone between cold polar and warm tropical air masses, generally occurring in the middle latitudes.

Polar navigation—Navigating becomes difficult in the polar regions because of the convergence of the meridians and the proximity of the magnetic north pole. Charts must be used with a so-called grid overlay in order to permit the pilot to fly a straight line from departure to destination. This system translates the rapidly changing true course measurements to a constant grid course.

Pop—Slang for papa, the phonetic term for P.

Position lights—Navigation lights.

Position report—Reports relating to an aircraft's progress given to ATC by the pilot when passing a geographical or nav-aid fix. It should include aircraft identification, time over the fix and altitude. It must also include the ETA for the next fix when the aircraft is not in radar contact.

Position symbol—A symbol generated by a computer and displayed on the radar screen at ATC facilities to indicate the tracking mode being used.

Position control—The term means that all traffic, IFR and VFR, is being controlled by and must maintain contact with ATC.

Positive control area—Airspace from 18,000 feet msl up to and including FL 600 (60,000 feet msl) above the 48 contiguous states and most of Alaska. In parts of Alaska its base is at 24,000 feet.

Power—Thrust produced by means other than gravity.

Power approach—An approach using partial power, resulting in a flatter angle of glide and better control over the aircraft than is possible in a power-off approach.

Power dive—A dive, at a steep angle, with power on.

Power glide—A gradual descent with power on.

Power landing—A landing at which partial power is used during final flare and until actual touchdown. It is used primarily by high-performance aircraft to improve control and insure an easier go-around if such should suddenly become necessary.

Power loading—The ratio of the certificated gross weight or the gross weight used at any given time to the rated horsepower of the engine(s).

Power-off glide—A glide with the engine at idle.

Power-off stalling speed—The speed at which the aircraft will stall with its power off. These speeds vary with flap and gear settings.

Power-on stalling speed—The speed at which the aircraft will stall, using partial or full power, by increasing the angle of attack beyond the point at which it maintains adequate lift generation. The speeds vary with gear position and flap settings, if any.

Powerplant—Engine(s).

Power settling—A phenomenon which can occur when a helicopter descends into its own downwash, effectively stalling part or all of the rotor.

pph—Pounds per hour. Used in measuring fuel flow.

Precession—The tendency of a gyroscope to gradually become unreliable, primarily due to friction. Usually used with reference to the directional gyro which tends to wander from the heading to which it has been set and must periodically be adjusted in accordance with the magnetic compass.

Precipitation—Visible moisture in the atmosphere, condensed to a point at which it will fall in the form of rain, drizzle, snow, hail, etc.

Precision approach—An instrument approach using ground-based electronic glide-slope guidance. An ILS or PAR approach.

Precision approach radar—A radar designed to enable the ground controller to provide the pilot with precise range, azimuth and altitude information throughout the final approach.

Precision turn—A turn made to a specific heading or covering a predetermined number of degrees while maintaining altitude. A training maneuver.

Preferential routes—The term refers to a route structure entered into computer memories at various ARTCCs in order to expedite handoffs from one sector or ARTCC to another. They are divided into preferential departure routes (PDR), preferential arrival routes (PAR) and preferential departure and arrival routes (PDAR). In many (though not all) instances PDRs are included is SIDs, PARs in STARs.

Preferred IFR routes—A route system established in heavy traffic areas designed to increase the efficiency and capacity of the ATC system. They may extend through one or several ARTCC areas. Preferred IFR

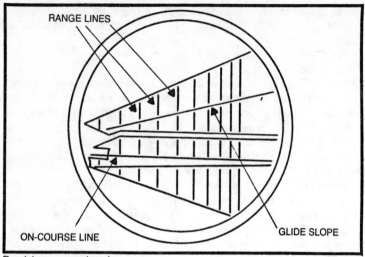

RANGE LINES

GLIDE SLOPE

ON-COURSE LINE

Precision approach radar.

routes are listed in AIM Part 3, and are coordinated with SIDs and STARs. Pilots are urged to file preferred IFR routes and, barring weather avoidance procedures or other considerations, ATC will issue clearances based on this route structure. A pilot may refuse acceptance of such a clearance if he deems it unsafe or uneconomical.

Preferred routes—Preferred IFR routes.

Preflight—Used as a verb it means to conduct a preflight inspection and/or to prepare the aircraft for flight.

Preflight inspection—A visual inspection of the aircraft before flight intended to discover anything that might result in problems during the flight. It should always include an oil check and a visual check of the level of fuel in the tanks. During winter months it should include the removal of frozen moisture from the airfoils and snow or ice accumulations in the gear assemblies, wheel wells, etc.

Preignition—The burning of the fuel-oil mixture in the combustion chamber before the spark plugs have had an opportunity to fire. It is caused by a hot spot in the engine such as overheated carbon deposits. It usually follows excessive overheating of the engine and results in erratic power output, increased overheating, and, in extreme cases, severe damage to the engine.

Pressure altitude—The indicated altitude when the altimeter is set to 29.92 in Hg. It is equal to the true altitude under standard pressure and temperature conditions and is used to calculate true altitude and density altitude. It is always used when flying at 18,000 feet msl or above.

Pressure gradient—The decrease in atmospheric pressure per unit of horizontal distance in the direction in which the pressure decreases most rapidly. It is perpendicular to isobars.

Pressure pattern flying—Flying an apparently circuitous route to take advantage of favorable winds resulting from prevailing pressure patterns.

Pressure relief valve—A valve which is part of the propane-tank system on hot-air balloons, designed to open when pressure increases to the danger point, closing again automatically when normal pressure has been restored.

Pressure suit—A suit used by pilots, usually military, designed to counteract the effects of excessive G loadings.

Pressurization—A system of maintaining given degrees of atmospheric pressure in the cabin of an aircraft. Pressurization amounts are measured in psi, and a given number of psi will maintain sea level pressure to a given altitude. (See chart). In pressurized piston-engine aircraft the pressurization is produced by a turbocharger. In turbine aircraft special pressurization equipment must be installed.

Pressurization system—The equipment on an aircraft to maintain reasonable pressure levels at high altitudes.

Pressurized aircraft—An aircraft equipped to maintain a pressure differential between the cabin and the outside.

Pressurized Centurion—A pressurized single-engine high-performance aircraft manufactured by Cessna Aircraft Company.

Pressurized Navajo—A pressurized cabin-class piston-twin manufactured by Piper Aircraft Corporation.

Pressurized Skymaster—A pressurized push-pull piston twin manufactured by Cessna Aircraft Company.

Prevailing visibility—The visibility in terms of distance at a given time and place.

Prevailing westerlies—The circulation pattern of the predominantly westerly winds between 30 and 60 degrees N latitude.

Prevailing wind—The wind direction most frequently observed at a given area. Runways are usually aligned to take advantage of the prevailing winds.

Preventive maintenance—Simple maintenance, repair and cleaning operations or replacement of standard parts not involving major disassembly and assembly operations.

PREFERRED ROUTES	
NEW ORLEANS METRO AREA	
ATLANTA	PCU V70 MVC V20 TYRONE (L17, 18, 20)
BIRMINGHAM	PCU V455 MEI V154 V209 BWA DRCT BHM/LOM (L-17, L-18, L-14)
DALLAS	WALKER V-114N AEX V-114 GGG V-94 V-477E FORNEY (L-17, 13)
DULLES	PCU V70 MVC V20 SBV V39 GVE V39E BRANDY (L-17, 18, 20, 22)
HOUSTON	TBD V20S LFT V20 LCH V222N MONUMENT (L-17)
KENNEDY	PCU V70 MVC V20 V213 ENO V44 BEECHWOOD (L-17, 18, 20, 22, 24)
MEMPHIS	PCU V9E MCB V9 GRW V9E INDEPENDENCE (L-14)
NEWARK	PCU V70 MVC V20 V213 ENO V29 V433 ROCKY HILL (L-17, 18, 20, 22, 24)
WASHINGTON	PCU V70 MVC V20 SBV V39 GVE V16 IRONSIDES (L-17, 18,20,22)

Preferred routes (courtesy of FAA).

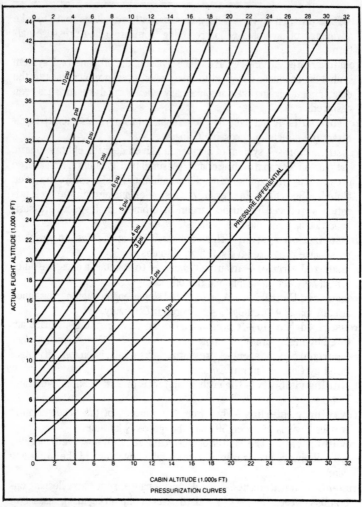

Pressurization curves.

Primary radar—Radar which transmits a pulsed signal which is reflected from objects and received back at the ground station, without the reflecting objects being in any way involved in augmenting or retransmitting the signal.

Primary target—A radar return from an aircraft which is not transponder equipped. Usually weak, it tends to get lost among the transponder-augmented returns.

Primer—A plunger-type of cockpit control which releases fuel directly into the cylinders, improving smooth starts especially under cold-weather conditions.

Cessna Pressurized Centurian (courtesy of Cessna).

Private license—A pilot's license which permits the pilot to carry passengers, but not for remuneration.

Private pilot—A pilot who has been issued a private license by the FAA. He must be 17 years old to be rated to fly powered aircraft (16 years for gliders), must speak, read and understand English, must have at least a third-class medical certificate, and must have accumulated a given amount of flight time (solo and dual), must have passed a written examination and a practical flight test with an authorized examiner and have demonstrated his flying and navigating capability.

Procedure turn—A turn at a definite fix, at or above a definite altitude and within a defined amount of air space, usually a part of an instrument approach for the purpose of maneuvering the aircraft into the right position, heading and altitude for the final approach. A procedure turn consists of a 45-degree turn to the left (occasionally right) followed by a 225-degree turn to the right (left) which should terminate at the same place at which the first turn was started.

Pressurized Navajo interior (courtesy of Piper).

263

Cessna Pressurized Skymaster (courtesy of Cessna).

Procedure turn inbound—That point at which the procedure-turn maneuver has been completed and the aircraft should be established on the final (or intermediate) segment of the approach. The phrase *procedure turn inbound* may be used by the pilot in reporting his position to ATC, or by ATC for separation purposes.

Profile descent—A procedure affecting primarily jet aircraft, designed to minimize noise as well as fuel consumtion during the descent and approach portion of the flight. While different descent profiles have been established by the FAA for different airports, the basic principle involves keeping the aircraft above FL 200 (20,000 feet msl) as long as possible (especially when delays are anticipated) and then to execute the entire descent at idle thrust all the way down to the final approach path. Airspeed assignments to speeds below 210 knots are to be minimized, and traffic flow regulated to avoid delaying maneuvers at low altitudes and speeds.

PROG—Prognosis.

PROG—Progress.

Prognostic chart—Surface-weather forecast chart.

Progress report—Position report.

Prohibited area—Airspace within certain geographic limits within which civil flight is prohibited. Prohibited areas may be used by the military for purposes which could endanger aircraft in the area or they may simply be portions of airspace above which aircraft flight is undesirable. The White House and any permanent or temporary residence of the President of the U.S. are prohibited areas, as is, for instance, the area covering the White Sands Proving Grounds and Los Alamos Scientific Laboratory in New Mexico.

ProLine—A family of sophisticated avionics designed for large high-performance aircraft and manufactured by Collins.

Prop—Propeller.

Prop—Used as a verb it means starting an aircraft by manually swinging the propeller.

Propane—Liquefied petroleum gas (LPG), the fuel used to heat the air in hot-air balloons.

Propeller—A device consisting of two or more air-foil-shaped blades which is designed to convert the turning force of the engine into thrust. Propellers may be *fixed-pitch*, meaning that the pitch of the blades cannot be varied; or *variable pitch* (also known as *constant speed*) in which case the pitch of the blades can be varied by the pilot in flight. Variable-pitch propellers are more efficient. The rotation speed of a propeller is limited to the speed at which the propeller-blade tips approach the speed of sound.

Propeller reverse—The ability to reverse the blade pitch of a propeller, resulting in negative thrust. Usually available only on turboprop aircraft.

Propeller spinner—The central shaft of a propeller to which the blades are attached. It is usually covered by a streamlined cone and this cone itself is often referred to as the spinner.

Propeller wash—The fast-moving disturbed airstream caused by a rotating propeller. It tends to pick up and blow back sand and small rocks which could damage aircraft or vehicles located behind an airplane with a fast-turning propeller.

Propping an airplane—Starting the engine by manually swinging the propeller. The only means of starting an airplane with no electrical system or one in which the battery is dead. In cold weather the propeller should be swung manually a few times (with the ignition system turned off) in order to loosen the oil in the engine.

Propulsion—The action or process of propelling. Thrust.

PRST—Persist.

PSBL—Possible.

PSBT—Pilot self-briefing terminal.

PSG—Passing; passage.

psi—Pounds per square inch. A measure of pressure.

PSN—Position.

PSR—Packed snow on runway.

PST—Pacific standard time.

Psychrometer—A hygrometer used for calculating relative humidity by comparing temperatures on a dry-bulb thermometer and a wet-bulb thermometer (one covered with a wet cloth), indicating rate of evaporation.

PT—Procedure turn.

PTLY—Partly.

PTN—Position.

Published—With reference to aircraft maneuvers, procedures which have been established and published by ATC.

Published approach—Standard instrument approach at any given location.

Published route—A route for which an IFR altitude has been established and published, such as federal airways, jet routes, etc.

Aerospatiale Puma. (courtesy of Aerospatiale).

Pulse instruments—Electronic instruments which emit pulsed (intermittent) transmissions. Radar is a pulse instrument, as are DMEs and radar altimeters.

Puma—A 16-place twin-turbine helicopter manufactured by Aerospatiale in France.

Pure jet—A turbojet engine or aircraft, not equipped with a fan or bypass function. Pure jets are noisier than fanjets, but are currently more efficient at very high altitudes (above 35,000 feet msl).

PVL—Prevail.

PVT—Private.

PWI—Proximity warning indicator.

PWR—Power.

Pylon—Any prominent mark or feature on the ground used as a fix in executing precision maneuvers.

Pylon eight—A training maneuver in which the aircraft describes a symmetrical figure-eight path over the ground, using two selected points (pylons) as the hub of each turn. It requires careful control of airspeed, rate of bank and turn, and correction for wind drift.

Pylon race—An air race using two pylons as the turning points in a racetrack-shaped flight path.

Q—Quebec (phonetic alphabet).

Q—Squall (in sequence reports).

Q altimeter setting—Several different kinds of altimeter settings, such as QFE and QNH. (The Q is an abbreviation left over from years-ago use with Morse code.)

QB—Quiet Birdmen, a primarily social organization of pilots.

Q-code—The code used in NOTAMs, always preceded by the letter Q.

QFE—An altimeter setting in which the altimeter is set in such a way that it will read zero altitude upon touchdown at the selected airport. It reduces the need for complicated arithmetic in the cockpit during an instrument approach and is in common use by some U.S. and by many European airlines. When a QFE altimeter setting is used, a second altimeter must be in the aircraft, set to the conventional setting based on barometric pressure.

QFLOW—Quota flow control.

QNH—The conventional altimeter setting based on the prevailing barometric pressure and reading altitude in feet msl.

QT—Quart.

QUAD—Quadrant.

Quadrant—One quarter of a circle. Also an instrument for measuring altitudes, commonly consisting of a graduated arc of 90 degrees with an index or vernier, usually having a plumb line or spirit level for fixing the vertical or horizontal direction. It is part of the instrumentation used in celestial navigation.

Quebec—In aviation-radio phraseology the term used for the letter Q. (Pronounced Kebeck.)

Queen Air—A cabin-class piston-engine twin aircraft manufactured by Beech Aircraft Corporation.

Beech Queen Air (courtesy of Beech).

Quick drain valve—A valve installed on the outside of the aircraft and connected to a fuel tank or fuel strainer through which water and sediment is collected and can be emptied during the preflight inspection.

Quick look—A feature of the ATC radar system which permits the controller to briefly display full data blocks of tracked aircraft from other control positions.

Quota flow control—An ATC procedure which limits the number of aircraft which may approach or enter a certain (terminal) area, designed to minimize congestion or saturation of that area or sector.

R—Ceiling measured by radar (in sequence reports).

R—Rain (in sequence reports).

R—Romeo (phonetic alphabet).

RAD—Radial.

Radair Division—Manufacturers of avionics equipment (3520 Pan American Freeway NE, Albuquerque, NM 87107.)

RADAR—Radio detection and ranging.

Radar—A family of electronic devices which use the interval between the time a radio signal is transmitted and the time the same signal, reflected off an object, is received back, to locate that reflecting object. The reflecting object can be an aircraft, precipitation, a ground-based feature, etc. By using combinations of special-use antennas, computers, different types of receivers and tarnsmitters, radar echoes are transformed into useful information for pilots and ATC controllers.

Radar advisory—Advice or information based on radar observations.

Radar air traffic control facility—(RATCF)—An ATC facility at a Navy or Marine Air Station, using PAR in combination with communication equipment to provide approach-control service to aircraft operating in its area of jurisdiction.

Radar altimeter—An altimeter using radar technology to accurately measure the distance between the aircraft and the ground. Usually usable only up to a distance of 2,500 feet. Also called *radio altimeter*.

Radar altimeters comparison chart.

MANUFACTURER	MODEL	PRICE estimated	Volts input DC	AC 400 Hz	Type cont. wave	pulse	Display numeric	pointer	dial	linear	logarithmic	Altitude range (feet)	Accuracy percent	Pitch limits degrees +/-	Roll limits degrees +/-	DH signal visual	audio	Wheels-up warning	Units	Weight lbs.	REMARKS
BONZER	MINI MARK	995	14 28									+80–1,000	7	30	30	OPT	OPT		2	2.5	
KING	KRA 10	2,200	28									+20–2,500	5-7	20	20				1	2	
BONZER	MARK 10X	2,295	14 28									+40–2,500	5-7	30	30				3	4.5	
SPERRY	AA-100A	2,990	14 28									+40–2,500	5-10	45	45				2	6.4	SERVOED INDICATOR
GOULD	HRA-100	4,000	N.A.									0–2,500	5-7	30	30				2	7.2	
KING	KRA 405	5,040	28									0–2,000	3-5	20	25				2	11	HELICOPTER VERSION AVAILABLE
COLLINS	ALT-50	5,883	28								*	0–2,000	2-5	40	50				4	9.4	*DIGITAL INDICATOR OPTION
SPERRY	AA-215	6,350	28									0–2,500	5-7	45	45				3	10.5	
SPERRY	AA-235	6,790	28									0–1,500	5-7	45	45				3	10.5	HELICOPTER VERSION AVAILABLE
COLLINS	ALT 55	6,915	28								*	0–2,500	2-5	40	50				4	7.9	*DIGITAL INDICATOR OPTION
COLLINS	AL 101	11,913 14,848		115								-20–2,500	2-5	40	50				2	22	VERTICAL SCALE TAPE DISPLAY
BENDIX AVIONICS	ALA-51A	13,576	N.A.									-20–2,500	2-5	30	30				3	18	RT/INDICATOR/2 ANTENNAS AEINC 552/552A FM/FM-CW
LITTON AMECOM DIV.	506	14,000	N.A.									-20–2,500	2-5	30	30				3	19.4	

Radar approach—An instrument using ASR or PAR radar.

Radar approach control—(RAPCON)—See radar air traffic control facility.

Radar arrival—An aircraft is being vectored to the final approach course or to the traffic pattern for a visual approach. See *visual approach*.

Radar beacon—The signal emitted by a secondary radar answered by a transponder or altitude encoder in an aircraft.

Radar contact—An ATC phrase to inform the pilot that the controller has him identified on radar.

Radar contact lost— An ATC phrase to notify the pilot that radar identification has been lost (at least temporarily).

Radar environment—That portion of the airspace in which radar service is being provided.

Radar flight following—The tracking of an aircraft which has been identified on radar, along its route of flight.

Radar identification—The process of identifying a specific radar return as being generated by a specific aircraft. If the aircraft is transponder equipped the controller can effect identification by asking the pilot to "squawk ident" which causes the radar return to brighten on the radar display. In order to positively identify an aircraft which is not transponder equipped, the controller will usually ask the pilot to execute a number of turns which then show up on the radar display.

Radar identified aircraft—An aircraft which has been positively identified on radar and the position of which is therefore known to the controller.

Radar monitoring— Providing radar service.

Radar navigation guidance—Providing radar service.

Radar point out—A phrase used among controllers in instances when one controller hands an aircraft over to another but intends to continue communication with that aircraft for coordination purposes.

Radar report—(RAREP)—Hourly reports from a weather bureau radar station covering storms and precipitation within approximately 100 miles of that station.

Radar route—A route flown by an aircraft being vectored. Navigational guidance and required altitudes are provided by ATC.

Radar safety advisory—Advisories issued by ATC to aircraft which, in the opinion of the controller, are getting dangerously close to the terrain or some sort of obstruction; or to advise pilots of a potential traffic conflict. Once having been advised by the pilot that he has taken corrective action or has the source of the danger in sight, such advisories are discontinued. Issuing such advisories is contingent on the ability of the controller to be aware of the potential danger, and once such an advisory has been issued it is the responsibility of the pilot(s) and not the controller to take the appropriate evasive action.

Radar separation—Separation of IFR aircraft based on radar observations.

Radar service—An all-encompassing term covering all services provided by ATC to pilots, using radar.

Radar service terminated—An ATC phrase used to inform the pilot that no further radar service is being provided to that particular aircraft.

Radar surveillance—Radar observations of a specific geographical area.

Radar traffic information service—See *radar advisories*.

Radar vector—Using radar to give navigational information and instructions to an aircraft.

Radial—Any of the 360 magnetic courses from a VOR, VORTAC or TACAN station, starting with zero at the magnetic north and increasing clockwise (90 degrees = East; 180 degrees = South; 270 degrees = West; 360 or zero degrees = North.)

Radial engine—A reciprocating engine in which the cylinders are arranged in a circle around a central crankshaft; Usually more cylinders are employed than in horizontally-opposed engines, providing an increase in power output. Radial engines are no longer used in modern light aircraft.

Radio altimeter—Radar altimeter.

Radio beacon—See *non-directional beacon (NDB)*.

Radio direction finding—See *direction finding (DF)*.

Radio facility—Any electronic nav aid.

Radio facilities chart—An aeronautical chart used primarily in IFR operations and showing all electronic nav aids, airways, distances between fixes and a world of other information, but no terrain features. Produced by the NOAA and by Jeppesen-Sanderson, Inc.

Radio fix—A position determined by the intersecting point of two or more radials from nav aids.

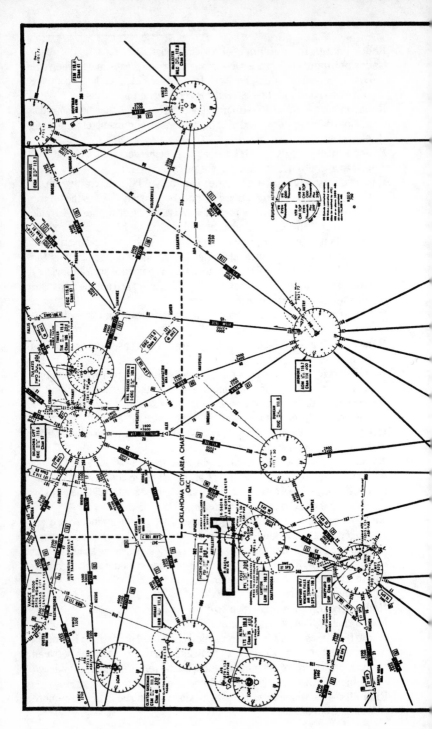

OKLAHOMA CITY AREA CHART
OKC

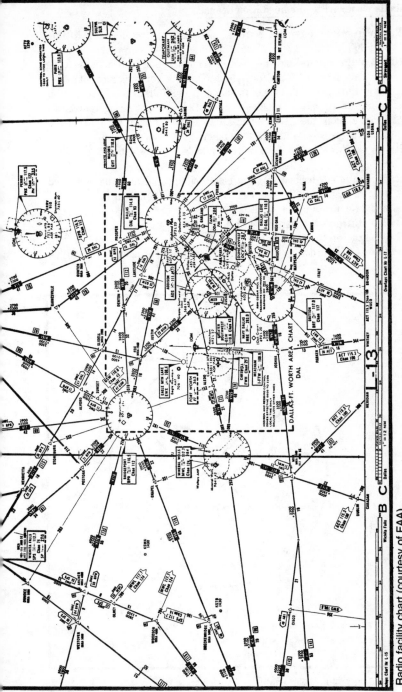

Radio facility chart (courtesy of FAA).

273

Radio frequencies—All electro-magnetic frequencies between 10 kHz and 300 MHz.

Radio magnetic indicator—(RMI)—A combination gyro compass (usually slaved) and VOR and/or ADF display. The compass card rotates to show the current heading of the aircraft; the VOR needle indicates the position of the aircraft relative to a selected radial or bearing from or to a given station; the ADF needle shows the position of the nose of the aircraft relative to the selected station.

Radio range—LF/MF radio range; a now largely obsolete aid to navigation.

Radiosonde—A balloon carrying electronic equipment to measure and transmit data about the upper air. It can be traced to measure the velocity and direction of upper winds.

Radio station—Standard broadcast station.

Radio station license—The license an aircraft owner must obtain from the FCC in order to legally operate the radio equipment aboard his aircraft.

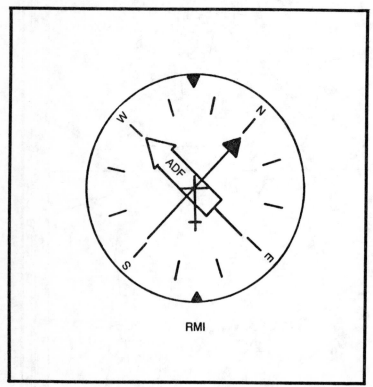

Radio magnetic indicator (RMI).

Radius—One half of the diameter of a circle.

Radome—A plastic cover for the antenna of airborne weather radar. Usually painted black and installed in the nose of multi-engine aircraft.

RAF—Royal Air Force.

RAIL—Runway alignment indicator lights.

Rain—Precipitation in the form of water droplets larger than those referred to as drizzle. The sequence-report symbol is R.

Rajay Industries—Manufacturers of turbochargers for reciprocating engines. (P.O.Box 207, Long Beach, CA 90801.)

Rallye—A family of single-engine aircraft manufactured by Aerospatiale in France.

Ram air—Air being forced into an orifice at a speed resulting from the motion of a vehicle. Mooney aircraft are equipped with a ram-air inlet to be used at the pilot's discretion when the air appears to be free of polluting particles. It produces a slight increase in engine-power output.

Ram-air pressure—Pitot pressure.

Ramp—The parking apron and adjacent areas on an airport, used for loading, unloading, tiedown.

Range—The distance an aircraft can travel on the available fuel. IFR range implies a 45-minute fuel reserve. No-reserve range implies the distance it is possible to fly until all tanks run dry.

Range—A radio or beacon system, as in VFR omni range.

Rangemaster—A high-performance single-engine aircraft produced by Navion. No longer in production.

Ranger—A high-performance single-engine aircraft manufactured by Mooney Aircraft Corporation.

RAPCON—Radar approach control.

RAREP—Radar report.

RATCC—Radar air traffic control center.

RATCF—Radar air traffic control facility.

Rate climb—A climb made at a constant rate requiring power and/or airspeed adjustments in order to maintain a constant fpm figure.

Rated horsepower—The maximum horsepower for which an engine has been certificated.

Rate of climb—The vertical distance traveled in terms of minutes, shown as fpm on the vertical-speed indicator.

Rate of climb indicator—Vertical-speed indicator (VSI).

Rate of descent—Opposite of rate of climb.

Rate of descent indicator—Vertical speed indicator (VSI).

Rating—A qualification of a pilot to fly a particular type of aircraft (jet-rated) or to fly IFR (instrument rated).

Raven Industries, Inc.—Manufacturers of hot-air balloons. (P.O.Box 1007, Sioux City, SD 75101.)

RBN—Radio beacon.

RCA Avionics Systems—Manufacturers of airborne weather radars and other high-performance avionics. (8500 Balboa Boulevard, Van Nuys, CA 91409.)

RCAG—Remote communication air/ground facility.

RCC—Rescue coordination center.

Reciprocal—The exact opposite bearing or radial, course or heading, by 180 degrees. To arrive at the correct reciprocal number of degrees, either add or deduct 180, depending on which results in a number of 360 degrees or less. (The reciprocal of 193 is 013; of 289 it is 109.)

RCH—Reach.

RCLM—Runway centerline marking.

RCLS—Runway centerline light system.

RCO—Remote communication outlet.

RCR—Runway condition reading.

RCV—Receive.

RCVG—Receiving.

RCVR—Receiver.

RDG—Ridge.

RDO—Radio.

Read—Phrase used in aviation-radio phraseology to mean *hear*. Such as in *I read you*.

Read back—A request by ATC asking the pilot to repeat a clearance or instruction.

Reading Air Show—Officially known as the National Maintenance and Operations Meeting, it is one of the biggest and most important annual general aviation events in the U.S. It usually takes place in the last week of May or the first week in June, attracts thousands of visitors from all over the country, and is often used by manufacturers to introduce new aircraft, avionics or other products. (Reading, PA is pronounced Redding.) (For information write P.O. Box 1201, Reading, PA 19603.) Warn-

Crowds at the Reading Air Show.

A Skymaster on floats at the Reading Air Show.

 ing: Hotel and motel rooms within a reasonable distance of Reading tend to be solidly booked a year in advance.

Ready for takeoff—Phrase used by pilots to inform the tower that they are ready to accept takeoff clearance.

Ready to copy—Phrase used by pilots in talking either to ground control or clearance delivery, to inform ATC that they are ready to listen to (and actually copy in writing, though this is not mandatory) the clearance for their IFR flight, the flight plan for which has been previously filed with ATC.

277

Ready to taxi—Phrase used by pilots to inform ground control that they are ready to leave the ramp or parking area in order to taxi out for takeoff. Often followed by a description of the type of flight: *Ready to taxi VFR Dallas; or ready to taxi IFR Albuquerque.*

Receiver—A radio instrument designed to receive either com or nav transmissions.

Receiving controller—Controller receiving the control over an IFR aircraft from another controller or facility.

Receiving facility—ATC facility receiving control over an IFR aircraft from another facility or controller.

Reciprocating engine—An engine which converts energy from fuel to mechanical motion by driving pistons back and forth. It primarily consists of cylinders, pistons, connecting rod and crankshaft, operating in a continuous four-stroke cycle.

RECONSTR—Reconstruction.

Recorder (flight data and/or cockpit voice)—(FDR or CVR)—Hermetically sealed recorders designed to keep a record of changes in aircraft attitude, control inputs and other flight data; or of what is being said in the cockpit by members of the crew. Usually installed in the tail of the aircraft, it is built to survive a crash and thus to serve investigators by giving them information as to the possible cause of an accident.

Recover—To return to normal flight after an intentional or unintentional spin or other unusual maneuver.

Recover—Replacing the fabric on a fabric-covered aircraft or section of an aircraft.

Redline speed—Never-exceed speed.

Redline temperature—Never-exceed air temperature in a hot-air balloon.

Reduce speed to (knots)—ATC request for a pilot to slow to a given speed, usually for purposes of separation.

Region (FAA)—The FAA is divided into seven regions: Eastern (Hq: Kennedy International Airport, Jamaica, NY.), Southern (Hq: Atlanta, GA), Central (Hq: Kansas City, MO), Southwest (Hq: Fort Worth, TX), Western (Hq: Los Angeles, CA), Alaska (Hq: Anchorage, AK), Hawaii (Hq: Honululu, HI). See also *General Aviation District Offices*.

Regional forecast—A weather forecast of large-scale phenomena covering several states, good for the next 24 hours.

Registration certificate—A certificate issued by the FAA, registering an aircraft to its owner. It must be carried aboard the aircraft at all times.

Registration mark—The national identification numbers and letters assigned to all civil aircraft.

Registration number—Registration mark.

Regulations—FARs.

REIL—Runway end identification lights.

Relative bearing—Bearing relative to a given fix.

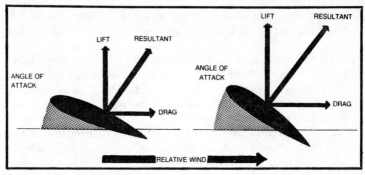

Relative wind (courtesy of FAA).

Relative humidity—The ratio of the amount of water vapor present in the air to the maximum amount it could hold at the prevailing temperature and at the same altitude. It is measured in *percent*.

Relative wind—The movement of air relative to the movement of an airfoil. It is parallel to and in the opposite direction of the flight path of the airplane.

Release—A mechanical means installed in gliders and sailplanes to permit the soaring pilot to release the tow rope which attaches his aircraft to the towing vehicle or mechanism.

Release time—A departure time slot issued by ATC to a pilot under conditions of congestion, when it becomes necessary in order to separate him from other departing or arriving traffic.

Relief—The contour and topography of the ground, shown on aeronautical charts in terms of variations in color and numbers giving the elevations in feet msl at given points.

Relief tube—A rather primitive excuse for a urinal, installed in some aircraft.

Remote communication air/ground facility—(RCAG)—An unmanned communications receiver/transmitter operating usually on VHF and UHF frequencies and designed to expand the receiving/transmitting range of ARTCCs.

Remote mounted—Refers to avionics and other equipment, the major portions of which are mounted somewhere other than behind the instrument panel. See also *panel mounted*.

Reno Air Races—Annual air races, usually involving unlimited-class aircraft, T-6s, midget racers, and aerobatic pilots in competition. Normally held some time in September.

Repeat—In aviation-radio phraseology a phrase preceding a second (or third) transmission of the same message. An alternate phrase is: *I say again*.

Report—Phrase used by ATC to ask pilots to give specific information; such as: *Report passing Anton Chico VOR*.

Reporting point—An exact fix over which a position report is made. For IFR aircraft not in radar contact, some such reporting points are mandatory.

Request full route clearance—A phrase used by pilots asking ATC to read the entire route of flight into the flight-plan clearance. It should always be used when an already filed flight plan has been subsequently amended for one reason or another by the pilot, in order to avoid any possible misunderstanding or confusion.

Rescue coordination center—(RCC)—RCCs are operated by the Coast Guard and/or the Air Force and are designed and equipped to coordinate search and rescue operations.

RESTR—Restrict.

Restraint systems—Seatbelts and shoulder harnesses.

Restricted area—Airspace above a specific geographical location within which flight of civil aircraft is not prohibited, but is limited at the discretion of the controlling agency. Permission to use such airspace should always be obtained in advance by contacting the controlling agency or the appropriate FSS.

Restricted landing area—Private airport, not for public use.

Restricted radiotelephone operators permit—A permit issued by the FCC, authorizing the pilot to communicate by radio from his aircraft (which, when in use, technically becomes a broadcast station).

Resume normal navigation—Phrase used by ATC after vectoring an aircraft or when losing radar contact while vectoring an aircraft, to inform the pilot to navigate on his own without further assistance from ATC.

Retractable gear—Retractable landing gear.

Retractable landing gear—Landing gear which, after takeoff, can be retracted into wheelwells or other housings in order to reduce drag. On most such aircraft all wheels retract fully. On some only the main gear retracts; and on some (the Bellanca Vikings) gears retract only partially.

Retreating blade stall—The stalling of a helicopter blade when the angle of attack of the retreating blade becomes too steep to generate lift. It tends to occur at high forward speeds.

Reverse thrust—Thrust in the opposite direction as that needed to propel the aircraft. Available primarily only on turbine-powered aircraft and used to reduce the landing roll. Extremely useful on icy runways when normal braking action becomes ineffective.

Reynolds number—Named after the English engineer Osborne Reynolds, it is an arbitrary number which relates to the flow of air over an airfoil (laminar or turbulent) and the point at which one turns into the other. The typical Reynolds number for a light aircraft is 3,000,000.

RGD—Ragged.

RGN—Region.

RGT—Right.

RHP—Rated horsepower.

Rhumbline—A line on the surface of the earth that intersects all meridians at equal oblique angles. It is a spiral coiling around the poles but never reaching them.

Ribs—Stiffeners inside wings and other aircraft components, fastened perpendicular to the surface of the skin.

Rich—With reference to engine operation, meaning a high percentage of fuel with relation to the amount of air. Rich mixture causes the engine to run cooler, uses an increased amount of fuel per hour, and tends to increase the chance of spark-plug fouling.

Ridge—An elongated area of high atmospheric pressure with the highest pressure along the centerline of the ridge.

Ridge soaring—Using the lift created by winds blowing at more or less right angles toward a mountain ridge to maintain altitude in a sailplane.

Rig—To modify or adjust control surfaces and/or trim tabs on the ground so the airplane will have more stable characteristics in the air. Generally done to reduce or eliminate tendencies for yaw or roll.

Right of way—When two aircraft are approaching head on, each shall alter its heading to the right. Or when two aircraft are converging at approximately the same altitude, the aircraft approaching from the right of the second aircraft has the right of way. The exceptions to these rules are: Airships, gliders and balloons have the right of way over heavier-than-air powered aircraft; balloons also have the right of way over gliders and, in fact, over all other aircraft including those having declared an emergency. Power-driven aircraft towing other aircraft or any other object have the right of way over all manner of power-driven aircraft.

Right traffic—A traffic pattern in which all turns are to the right, often established for certain airports or runways for purposes of noise abatement or to avoid higher terrain or obstacles. The existence of right or left traffic patterns at uncontrolled airports can be determined from the air by over-flying and observing the segmented circle.

Rigid rotor—Helicopter rotors with non-flexing blades, permitting flight at higher than normal forward speeds. Still primarily experimental, though used by Lockheed in its military Cheyenne helicopter.

Rime ice—Ice which forms as granular translucent or opaque chunks from the rapid freezing of supercooled water. It is softer and lighter than clear ice and gathers particularly on the leading edges of airfoils, causing deformation and increase in stall speed.

Rip panel—In balloons, a panel which can rapidly be opened to spill out hot air or gas, usually used at the moment of landing to reduce the chance of the balloon being dragged along by the wind.

Rivets—The customary means of bonding metal in aircraft construction. Rivets come in a wide variety of sizes and other characteristics. The three basic types are standard rivets with a rounded head, flush rivets which leave no protrusion on the surface of the metal being joined, and pop rivets, used primarily by homebuilders. Pop rivets do not require the use of a rivet gun.

Robinson helicopter. (courtesy of Robinson).

Robinson Helicopter Company piston helicopter specifications.

ROBINSON HELICOPTER COMPANY MODEL:	R22
ENGINE	L/O-320
hp	150 derated to 124
TBO (hours)	2,000
SEATS	2
RATE OF CLIMB (fpm)	1,200
SERVICE CEILING (ft)	14,000
HIGE (ft)	6,400
HOGE (ft)	4,500
MAX SPEED sea level (knots)	100
NORMAL CRUISE (knots)	94
Vne (knots)	102
FUEL CAPACITY (lbs)	120
RANGE w. full fuel (nm)	209
w. max payload (nm)	209
GROSS WEIGHT (lbs)	1,300
EMPTY WEIGHT (lbs)	780
PAYLOAD w. full fuel (lbs)	390
w. fuel for 100 nm (lbs)	453
FUEL w. max payload (lbs)	57
LENGTH/HEIGHT/WIDTH external (ft)	29/9/7
MAIN ROTOR number of blades	2
diameter (ft)	25.17
TAIL ROTOR number of blades	2
diameter (ft)	3.5
PRICE (1979 $s)	39,500 (approx.)

Rockwell turbo commander 690B (courtesy of Rockwell International)

RLA—Restricted landing area.

RMI—Radio magnetic indicator.

RMN—Remain.

RNAV—Area navigation.

RNAV approach—An instrument approach which requires area navigation instrumentation in the aircraft for navigational guidance.

RNG—Range.

RNWY—Runway.

Robinson Helicopter Company—Manufacturer of a low-cost two-seat piston helicopter. (Torrance Municipal Airport, Torrance, CA 90505.)

Rockwell International, General Aviation Division—Manufacturers of a wide variety of aircraft of the Commander family, ranging from

Rockwell Alpine and Gran Tourismo Commanders (courtesy of Rockwell International)

Rockwell International single-engine piston aircraft specifications.

ROCKWELL INTERNATIONAL MODEL:	COMMANDER GRAN TOURISMO 114A	COMMANDER ALPINE 112 TCA
ENGINE	L/IO-540-T4B5D	LTO-360-C1A6D
TBO (hours) (NUMBER OF SEATS)	NA (4)	NA (4)
PROPELLER	const. speed	const. speed
number of blades	3	2
LANDING GEAR	retractable	retractable
TAKEOFF ground roll (ft)	NA	
50′ obstacle (ft)	1,963	1,900
RATE OF CLIMB (fpm)	1,037	978
Vx (knots)	76	72
Vy (knots)	94	86
SERVICE CEILING (ft)	14,300	20,000 (max certif.)
MAX SPEED (knots)	166 (sea level)	170 (20,000 ft)
CRUISE SPEED 75% (knots)	165 (sea level)	163 (20,000 ft)
55% (knots)	137 (9,600 ft)	139 (20,000 ft)
RANGE w. res. 75% (nm) standard tanks	643	460
long-range tanks	NA	664
55% (nm) standard tanks	730	597
long-range tanks	NA	901
FUEL FLOW 75% (pph)	89	79.2
55% (pph)	66	51
STALL clean/dirty (knots)	61/56	58/54
LANDING 50′ obstacle (ft)	1,200	1,275
RAMP WEIGHT (lbs)	3,260	2,962
TAKEOFF/LANDING WEIGHT (lbs)	3,140	2,950
ZERO FUEL WEIGHT (lbs)	2,852	NA
USEFUL LOAD (lbs)	1,170	927
FUEL standard tanks (lbs)	408	288
long-range tanks (lbs)	NA	408
MAX PAYLOAD (lbs)	880	880
PAYLOAD w. full fuel, stand. tanks (lbs)	762	639
long-range tanks (lbs)	NA	519
FUEL w. max payload (lbs)	290	147
WING LOADING (lbs per sq ft)	21.4	18
WING AREA (sq ft)	152	163.8
LENGTH/HEIGHT/SPAN external (ft)	24.92/8.4/32.75	25.1/8.4/35.6
LENGTH/HEIGHT/WIDTH cabin (ft)	7.1/4.1/3.9	7.1/4.1/3.9
TURBOCHARGER make	NA	Garrett
PRICE (1979 $s)	95,010	91,900

STANDARD EQUIPMENT (both models): IFR PANEL; VHF NAV & COM w. OBI; AUDIO PANEL; MARKER; GLIDE SLOPE; TRANSPONDER; ENCODING ALTIMETER; EGT; CHT; ADF; DME; AUTOPILOT.

high-performance singles to corporate turboprops. (5001 Rockwell Avenue, Bethany, OK 73008.)

Rockwell International, Sabreliner Division—Manufacturers of the Sabreliner family of corporate jets. (6161 Aviation Drive, St. Louis, MO 63134.)

Rods—The sensitive receiving instruments in the human eye, placed around the cones. Rods are color blind, but are able to adapt to very low light conditions and provide us with night vision. Rods cannot see objects straight ahead.

Roger—In aviation-radio phraseology the term used to imply that the last transmission has been received and understood.

Rogue pilot—Pilots who ignore the regulations. Especially, pilots who fly IFR without maintaining contact with ATC when operating in controlled airspace.

Roll—Movement of the aircraft around its longitudinal axis. Also see *bank*.

Roll axis—Longitudinal axis.

Roll out—The completion of a turn and the return to level flight.

Rollout RVR—RVR measured by equipment located near the roll-out end of the runway.

Romeo—In aviation-radio phraseology the term used for the letter R.

RON—Rest over night. Remain over night.

Rotary-wing aircraft—Helicopter; gyrocopter; gyroglider.

Rotate—The moment prior to actual liftoff when the nosewheel is raised off the ground, but the mainwheels are still in contact with the runway surface.

Rockwell International twin-engine piston aircraft specifications.

ROCKWELL INTERNATIONAL MODEL:	SHRIKE	COMMANDER 700
ENGINES	L/IO-540-E1B5 (2)	L/TIO-540-R2AD (2)
TBO (hours)	NA	NA
PROPELLERS	const. speed, ff	const. speed, ff
number of blades	3	3
SEATS	6	7
TAKEOFF ground roll (ft)	NA	1,604
50′ obstacle (ft)	1,915	2,264
RATE OF CLIMB (fpm)	1,340 (1 eng. 266)	1,579 (1 eng. 261)
Vx (knots)	NA	90
Vxse (knots)	NA	98
Vy (knots)	103	108
Vyse (knots)	94	105
SERVICE CEILING (ft)	19,400 (1 eng. 6,500)	26,800 (1 eng. 10,600)
MAX SPEED (knots)	187 (sea level)	221 (17,000 ft)
CRUISE SPEED (knots)	75%:176 (6,500 ft)	75%:212 (21,500 ft)
economy (knots)	45%:142 (15,000 ft)	45%:155 (18,000 ft)
RANGE w. res. (nm)	75%:812 (6,500 ft)	75%:845 (15,500 ft)
economy (knots)	45%:976 (15,000 ft)	45%:1,202 (18,000 ft)
FUEL FLOW per engine (pph)	75%:103	75%:128
economy (pph)	45%:69	45%:67
STALL clean/dirty (knots)	68/59	86/68
LANDING ground roll (ft)	NA	1,089
50′ obstacle (ft)	3,150	2,154
ACCELERATE-STOP DISTANCE (ft)	NA	3,695
RAMP WEIGHT (lbs)	6,750	6,987
TAKEOFF/LANDING WEIGHT (lbs)	6,750	6,947/6,600
ZERO FUEL WEIGHT (lbs)	6,750	NA
USEFUL LOAD (lbs)	2,115	2,283
FUEL CAPACITY (lbs)	936	1,248
MAX PAYLOAD (lbs)	2,115	1,890
PAYLOAD w. full fuel (lbs)	1,179	1,035
FUEL w. max payload (lbs)	not possible	393
WING LOADING (lbs per sq ft)	26.47	34.7
WING AREA (sq ft)	255	200
LENGTH/HEIGHT/SPAN external (ft)	36.81/14.5/49	38.2/13.3/42.4
LENGTH/HEIGHT/WIDTH cabin (ft)	NA	16.4/4.7/4.7
TURBOCHARGER make/type	NA	Garrett/TH08A69
PRESSURIZATION (psi)	NA	5.5
PRICE (1979 $s)	NA	NA

Rockwell International turboprop aircraft specifications.

ROCKWELL INTERNATIONAL MODEL:	EXECUTIVE I	EXECUTIVE II
ENGINES	GA/TPE 331-5-251	GA/TPE 331-5-251
shp each	718 flat rated	718 flat rated
TBO (hours)	NA	NA
SEATS crew + passengers	2 + 6	2 + 6
TAKEOFF ground roll (ft)	1,458	1,458
50′ obstacle (ft)	2,259	2,259
RATE OF CLIMB (fpm)	2,821 (1 eng. 878)	2,821 (1 eng. 878)
Vxse (knots)	109	109
Vyse (knots)	115	115
Vmca (knots)	86	86
SERVICE CEILING (ft)	32,800 (1 eng. 19,600)	32,800 (1 eng. 19,600)
ACCELERATE-STOP DISTANCE (ft)	3,450	3,450
MAX SPEED (knots)	284 (17,500 ft)	284 (17,500 ft)
BEST CRUISE SPEED (knots)	283 (20,000 ft)	283 (20,000 ft)
LONG-RANGE CRUISE SPEED (knots)	250.3 (31,000 ft)	250.3 (31,000 ft)
RANGE w. res. best speed (knots)	1,036 (18,000 ft)	1,036 (18,000 ft)
long-range speed (knots)	1,467 (31,000 ft)	1,467 (31,000 ft)
FUEL FLOW per engine, best speed (pph)	294	294
long-range (pph)	178	178
STALL clean/dirty (knots)	82/77	82/77
LANDING ground roll (ft)	1,613	1,613
50′ obstacle (ft)	2,100	2,100
RAMP WEIGHT (lbs)	10,375	10,375
TAKEOFF/LANDING WEIGHT (lbs)	10,325/9,675	10,325/9,675
ZERO FUEL WEIGHT (lbs)	8,750	8,750
USEFUL LOAD (lbs)	3,642	4,180
FUEL CAPACITY (lbs)	2,573	2,573
MAX PAYLOAD (lbs)	2,017	2,555
PAYLOAD w. full fuel (lbs)	1,625	1,574
FUEL w. max payload (lbs)	2,573	1,625
WING LOADING (lbs per sq ft)	38.82	38.82
WING AREA (sq ft)	266	266
LENGTH/HEIGHT/SPAN external (ft)	44.35/14.95/46.67	44.35/14.95/46.67
LENGTH/HEIGHT/WIDTH cabin (ft)	14.25/4.47/4.02	14.25/4.47/4.02
PRESSURIZATION (psi)	5.2	5.2

Rotating beacon—The green and white rotating beacon denoting an airport, always operating after dark and usually when the airport is IFR. Also the red rotating beacon on the aircraft.

Rotor—The horizontally rotating air circulation under a mountain wave. Generally an area of severe turbulence.

Round robin—A cross-country flight in which the aircraft returns to its airport of departure without landing elsewhere in between.

Route—A defined path consisting of one or more courses flown in a horizontal plane.

Route segment—A definite portion of a route, identified by geographic or nav-aid fixes.

Route structure—Airway structure.

RPD—Rapid.

rpm—Revolutions per minute.

RPT—Repeat.

RR—LF/MF radio range.

RR—Railroad.

RSG—Rising.

RTE—Route.

RTRD—Retard.

RTRN—Return.

Rudder—The primary control surface attached to the vertical stabilizer, movement of which causes the tail of the aircraft to swing right or left. It controls yaw; movement around the vertical axis.

Rudder pedal—The foot-operated pedals which operate the rudder; also used to steer tricycle-gear aircraft on the ground. Most rudder pedals include the brake mechanism which is activated by depressing the toes and thus the top part of the pedals.

RUF—Rough.

Runup—A pretakeoff check of the performance of the engine and, in aircraft with controllable-pitch props, the operation of the propeller. The engine is run up to a given rpm and the pilot then switches from one set of magnetos to the other to make sure that each set works satisfactorily independent of the other. To exercise the propeller, he retracts the rpm knob or lever several times.

Runway—The portion of an airport designed for takeoffs and landings.

Runway condition reading—(RCR)—Determining the braking action under prevailing runway-surface conditions. Also see *braking action*.

Runway gradient—The difference in elevation between two ends or different portions of one and the same runway; measured in percent.

Runway in use—Active runway.

Rockwell International jet aircraft specifications.

ROCKWELL INTERNATIONAL MODEL:	SABRELINER 60	SABRELINER 75A	SABELINER 75A
ENGINES:	PW/JT12A-8	GA/TFE 731-3R-1D	GE/CF-700-2D-2
THRUST (pounds each)	3,300	3,700	4,500
SEATS crew + passengers	2 + 8	2 + 8	2 + 8
BALANCED FIELD LENGTH, ISA (ft)	5,300	5,150	4,380
ISA+20 (ft)	6,250	7,150	5,450
RATE OF CLIMB (fpm)	4,700 (1 eng. 1,100)	3,540 (1 eng. 950)	4,500 (1 eng. 1,050)
Vr (knots)	128	NA	133
V₂ (knots)	134	NA	137
SERVICE CEILING (ft)	45,000 (1 eng. 26,000)	45,000 (1 eng. NA)	45,000 (1 eng. 24,000)
Mmo	.81	85	80
ALTITUDE CHANGEOVER (FL)	215	240	215
Vmo (knots)	350	367	350
CRUISE hi speed (knots)	489 (FL 450)	513	489 (FL 450)
long range (knots)	402 (FL 390)	430 (FL 430)	408 (FL 390)
RANGE hi speed (nm)	1,492	NA	1,678
economy (nm)	NA	2,795	NA
FUEL FLOW per engine, hi speed (pph)	821.5	NA	880
long range (pph)	685.5	545	723
RAMP WEIGHT (lbs)	20,372	24,000	23,000
TAKEOFF/LANDING WEIGHT (lbs)	20,172/17,500	24,000/21,755	22,800/22,000
ZERO FUEL WEIGHT (lbs)	NA	16,250	NA
USEFUL LOAD (lbs)	9,122	11,050	9,800
FUEL CAPACITY (lbs)	7,122	8,825	7,380
FUEL w. max payload (lbs)	7,122	8,825	7,380
PAYLOAD w. max fuel (lbs)	1,600	1,825	2,020
WING LOADING (lbs per sq ft)	NA	63.2	NA
WING AREA (sq ft)	NA	380	NA
LENGTH/HEIGHT/SPAN external (ft)	48.3/16/44.6	46.9/16/50.4	47/17.25/44.43
LENGTH/HEIGHT/WIDTH cabin (ft)	19/5.5/5.2	19/5.5/5.2	19/6/5.2
PRESSURIZATION (psi)	8.8	8.8	8.8
EXTERIOR NOISE (EPNdB)	95/98.5	NA	90.7/100.2
PRICE (1979 $s)	NA	3,425,000	NA

Runway lights—Lights used to define the side limits of a runway, usually spaced 200 feet apart.

Runway markings—Centerlines, threshold markings and other painted signs designed to aid pilots in landing under VFR and/or IFR conditions.

Runway number—Each runway has a number based on its magnetic direction. Thus Runway 30 has a 300-degree alignment. Usually these

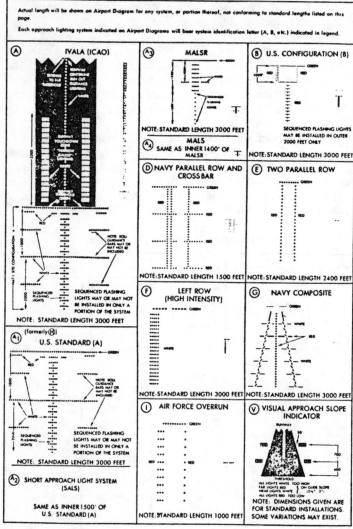

Approach lighting systems (courtesy of FAA).

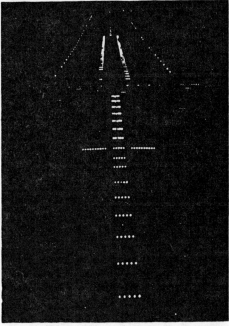

Instrument runway at night
(courtesy of FAA).

numbers are painted on the approach end of the runway surface, but not always. See also *parallel runways*.

Runway visibility—The horizontal distance at which a stationary observer near the end of the runway can see an ordinary light at night, or a dark object against the horizon or sky in the daytime. It is normally measured by a transmissometer and reported in statute miles or fractions thereof.

Runway visual range—(RVR)—The horizontal distance measured by a transmissometer and reported in hundreds of feet, representing the distance a pilot can see down the runway under low-visibility conditions. Different types of instrument-approach categories require different RVR minimums. See *Category I, II, III approach.*

RVR—Runway visual range.

RVV—Runway visibility value.

RW—Rain showers (in sequence reports).

S

S—Snow (in sequence reports).

S—South.

S—Special sequence report.

S—Straight-in (instrument approach charts).

Sabre—A family or corporate jet aircraft manufactured by the Sabreliner Division of Rockwell International.

SAC—Strategic Air Command.

SAE—Society of Automotive Engineers. See *aviation organizations.*

Safe Flight Instrument, Inc.—Manufacturers of angle-of-attack indicators. (20 New King Street, White Plains, NY 10602).

Safety factor—The safe limit of load in an airplane.

Safety factor—Any design component or procedure which assures a reasonable margin of safety.

Safety line—The drop line on a hot-air or gas balloon or a blimp or dirigible.

Saft America, Inc.—Distributors of nickel-cadmium batteries manufactured by Saft in France. (711 Industrial Boulevard, Valdosta GA 31601.)

Sailplane—A high-performance glider.

SALS—Short approach light system.

SAP—Soon as possible.

SAR—Search and rescue.

SAS—Stability augmentation system.

Saturated air—Air which contains all the water vapor it can at a given pressure and temperature. Equal to 100 percent humidity.

SAVASI—Simplified abbreviated visual approach slope indicator.

Say again—In aviation-radio phraseology the term used to mean, *please repeat your last transmission.*

Say altitude—ATC term requesting the pilot to state his current altitude.

Rockwell Sabre 60 (courtesy of Rockwell International).

A sailplane aloft.

Sailplanes on the ground.

Say heading—ATC term requesting the pilot to report the actual heading of the aircraft.

Scattered clouds—Sky cover indicating that clouds cover between 10 and 60 percent of the sky.

Scheibe Flugzeugbau GmbH—Manufacturers of high-performance sailplanes (August-Pfaltz Strasse 23, 806 Dachau bei Muenchen, West Germany. U.S. Distributor: Graham Thompson Ltd., 3200 Airport Avenue, Santa Monica, CA 90405.)

Schleicher Segelflugzeugbau—Manufacturers of high-performance sailplanes. (U.S. Distributor: Schleicher Sailplanes, P.O.Box 118, Port Matilda, PA 16870.)

Schweizer Aircraft Corporation—Leading U.S. manufacturer of gliders and sailplanes. (P.O.Box 147, Elmira, NY 14902.)

Scope—The face of the cathode-ray tube displaying radar returns.

Scott Aviation, Inc.—Manufacturers of oxygen equipment for aircraft. (225 Erie Street, Lancaster, NY 10486.)

SCTD—Scattered.

SCTR—Sector.

Scud—Small, low, wind-driven clouds.

SDF—Simplified directional facility.

Sea breeze—The movement of air from sea to land during the day. The opposite of the land breeze at night.

Sea breeze front—The line of convergence between warm inland air and the moist cool air from the ocean.

Sea-level conditions—Conditions affected by altitude and adjusted to sea level.

Seaplane—An aircraft with floats instead of wheels or a boat-like hull, operating to and from water.

Search and rescue—A service providing all needed functions and aid to locate missing aircraft and assist downed aircraft. It involves cooperation between the FAA, Air Force, Coast Guard, Civil Air Patrol and various state and local agencies.

Seat-of-the-pants—Flying by reference to the kinesthetic sense. It is unreliable and potentially dangerous under instrument conditions when visual reference to the ground or horizon is lost.

SEC—Second.

SEC—Section.

SEC—Sectional.

Second—One 60th of a minute of a circle. The symbol for a second is".

Secondary radar—Radar which involves both ground interrogator and airborne transponder signals. It is the type used in ATC radar beacons.

Sectional chart—An aeronautical chart of a section of the U.S. at a scale of 1:500,000 or approximately seven nm per inch. A total of 37 sectional charts cover the contiguous 48 states.

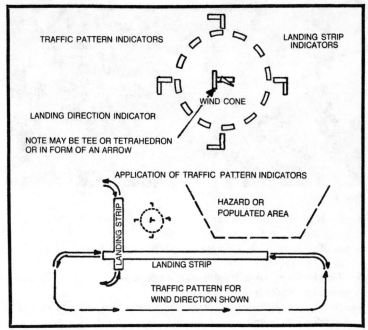

Segmented circle (courtesy of FAA).

Sector—A portion of the total air space controlled by a given ARTCC. Each sector has its own discrete frequency for communication between pilot and controller.

Sector frequency—The frequency assigned to a sector of an ARTCC.

See and avoid—See and be seen.

See and be seen—The concept of avoiding midair collisions by placing the responsibility for maintaining safe separation on the pilot. In VFR conditions, pilots, regardless whether flying VFR or IFR, are responsible for maintaining a visual alert for other traffic.

Segmented circle—A visual indication at uncontrolled airports showing the pattern direction for each runway. Usually associated with a windsock and/or tetrahedron.

Segments of an instrument approach procedure—Instrument approach procedures are divided into the *intial approach segment,* the *intermediate approach segment,* the *final approach* and the *missed approach procedure.*

SEL—Single engine land. A pilot rating.

Self-launched glider—A hang glider capable of being launched by the pilot using his legs.

Self-launched sailplane—Motorglider. A glider or sailplane equipped with an engine of some kind.

Piper Seminole (courtesy of Piper).

Semco Balloon, Inc.—Manufacturers of hot-air balloons. (Rte. 3, Box 514, Aerodrome Way, Griffin, GA 30223.)

Seminole—A light-light twin engine aircraft manufactured by Piper Aircraft Corporation.

Semi-rigid rotor—A helicopter rotor-blade assembly in which flexible blades are rigidly connected to the rotor hub.

Seneca II—A turbocharged twin-engine aircraft manufactured by Piper Aircraft Corporation.

Separation—Spacing of aircraft to achieve safe and orderly movement.

Separation minimums—The minimum distances in the longitudinal, lateral and vertical dimensions by which aircraft are spaced when operating under ATC control in IFR conditions.

Sequence report—An hourly report of observed, measured or estimated weather conditions transmitted over the national teletype system with coded information in a specific sequence and with all reporting stations in a given sequence. Portions of it are broadcast by FSSs at 15 minutes past the hour. When drastic changes in weather conditions occur, special reports are issued as needed (labeled S).

Service ceiling—The highest altitude at which a given aircraft can continue to climb at 100 fpm.

SES—Single engine sea. A pilot rating.

Servo mechanism—A mechanism which operates automatically when activated by some other mechanism or electronic signal.

Servo motor—The small electric motors which operate the control surfaces of an aircraft when they are activated by the autopilot.

Severe weather avoidance plan—(SWAP)—An ATC procedure in the New York area, designed to move traffic with minimum delays and detours when large portions of the air space are unusable due to thunderstorms or other severe weather.

Sextant—An instrument for measuring angular distances used in navigation to ascertain latitude and longitude by reference to celestial bodies.

Piper Seneca II (courtesy of Piper).

SFA—Single frequency approach.

SFC—Surface.

Shaft horsepower—The amount of horsepower delivered by the engine to the propeller or rotor shaft.

Shark—A piston helicopter manufactured by Enstrom Helicopter Corporation.

Shear—Windshear.

Shear line—The line between air masses moving in different directions or at different speeds.

SHF—Super high frequency.

SHFT—Shift.

SHLW—Shallow.

Shock strut—Oleo strut.

Shoran—Stands for short-range navigation. An electronic position-fixing system in which an airborne pulse transmitter triggers responses from two transponders of known location, providing a means of triangulation.

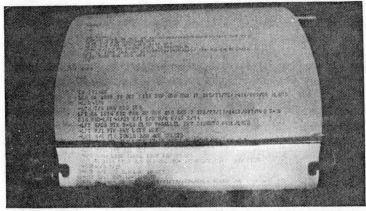

Sequence report coming off the teletype machine.

Rockwell Shrike Commander (courtesy of Rockwell International).

Short-field landing—A landing designed to result in the shortest possible landing roll. It is usually made with full flaps and considerable power, holding the nose higher than usual and arriving over the threshold barely above stall.

Short-field takeoff—A takeoff designed to get the aircraft airborne in the shortest possible distance. Normal technique calls for running the engine at full power while holding the aircraft in place with hard braking. Then let go of the brakes and feed in partial flaps, using whatever backpressure is needed to achieve liftoff at minimum flying speed. The aircraft, especially low-wing aircraft, may lift off at a speed slightly below normal stalling speed because of ground effect. Assuming no obstacle is in the way, the pilot should hold the aircraft in ground effect until sufficient speed has been gained to permit climbout.

Short range clearance—A clearance issued to an IFR flight which temporarily terminates at a fix short of the destination while ATC coordinates traffic and obtains the complete clearance.

Short takeoff and landing aircraft—(STOL)—Aircraft equipped with special high-lift devices which permit reduction in takeoff and landing roll.

Shower—Precipitation of short duration and limited in area with less gustiness than would be associated with a squall.

shp—Shaft horsepower.

Shrike Commander—One of the Commander family of piston twins, manufactured by the General Aviation Division of Rockwell International.

SHWR—Shower.

SID—Standard instrument departure.

Side load—The lateral loads exerted on the aircraft and especially the landing gear when the aircraft touches down in a slightly sideways direction, usually the result of crosswind.

Sideslip—Yaw.

Sidestep maneuver—A visual maneuver executed by a pilot at the end of a straight-in instrument approach, permitting him to land on a parallel runway. The distance between the two runways may not exceed 1,200 feet.

Sierra—In aviation-radio phraseology the term used for the letter S.

Sierra—A high-performance single-engine aircraft manufactured by Beech Aircraft Corporation.

SIFR—Special IFR. Also S/IFR.

SIGMET—Stands for significant meteorology. An inflight weather advisory covering weather phenomena of importance to all aircraft, such as lines of thunderstorms, tornadoes, large hail, severe turbulence, heavy icing, and widespread dust or sand storms which reduce visibility to less than two miles.

Signals—A variety of standard visual signals have been established, some used by line personnel in directing aircraft on the ground, others by the crews of downed aircraft to give information to overflying search aircraft. (See illustration under *emergency* and under *hand signals*.)

Sikorsky Aircraft Division—Manufacturers of turbine helicopters. (Stratford, CT 06602.)

Silver Instruments, Inc.—Manufacturers of electronic fuel-control systems. (1896 National Avenue, Hayward, CA 94545.)

Simplified directional facility—(SDF)—A non-precision-approach nav aid, similar to a localizer, except that it may be offset up to three degrees from the runway. It is often wider than a localizer and therefore less accurate.

Simulated flameout—(SFO)—A practice approach for jet aircraft executed at idle thrust all the way to the runway. Primarily a military training maneuver.

Simulator—Any device which simulates actual conditions, used for training purposes. Normally used with reference to simulated aircraft in which a pilot can execute virtually any maneuver that could be flown in a real airplane. Such flight simulators range from simple desk-top models, costing from $1,000 on up, to highly sophisticated machines with built-in motion and simulated pictorial displays representing the airport, runway,

Beechcraft Sierra 200 (courtesy of Beech).

Sikorsky Aircraft Division twin-turbine helicopters specifications.

SIKORSKY AIRCRAFT DIVISION MODEL:	S-76 SPIRIT	S-61N MARK II
ENGINES	AL/250-C30 (2)	GE/CT58-140 (2)
shp each	650	1,350
SEATS	14	22
RATE OF CLIMB max sea level	2,000	1,300
one engine out	NA	300
SERVICE CEILING (ft)	15,000	12,500
one engine out (ft)	4,800	3,400
HIGE (ft)	5,100	8,700
HOGE (ft)	1,300	3,800
MAX SPEED sea level (knots)	155	130
NORMAL CRUISE SPEED (knots)	155	121
Vne (knots)	161	131
FUEL CAPACITY (lbs)	1,876	4,382
RANGE w. full fuel (nm)	480.5	472
GROSS WEIGHT (lbs)	9,700	20,500
EMPTY WEIGHT (lbs)	5,273	12,510
USEFUL LOAD (lbs)	4,427	7,990
PAYLOAD w. full fuel (lbs)	2,151	3,208
LENGTH/HEIGHT/WIDTH external (ft)	52.5/14.5/7	73.1/17.6/19.9
cabin (ft)	8.1/4.4/6.3	31.9/6.3/6.5
MAIN ROTOR number of blades	4	5
diameter (ft)	44	62
TAIL ROTOR number of blades	4	5
diameter (ft)	8	10.6
IFR CERTIFICATION STATUS	pending	2-pilot
PRICE (1979 $s)	1,125,000	3,540,000

or other scenes under all imaginable weather conditions. Their prices run up into many millions.

Single direction route—Preferred IFR routes which are normally used only in one direction.

Single-engine aircraft—An aircraft propelled by a single powerplant. It can be a piston-engine, a turboprop or a jet.

Single frequency approach—(SFA)—An approach available primarily to single-piloted military jets, in which the pilot can stay on one and the same frequency throughout the approach.

Single-piloted aircraft—An aircraft with only one set of controls. Usually military.

Sink—Rate of descent. Used primarily in ballooning.

Sink—With reference to soaring, descending air in which the glider or sailplane loses more altitude than it would in still air.

SIR—Snow and ice on runway.

SKED—Schedule(d).

Skid—Lateral movement of an airplane toward the outside of a turn, caused by incorrect use of the rudder.

Skin—The outer covering of the airframe. It can be aluminum, fabric, wood, or fiberglass.

Beechcraft Skipper. (courtesy of Beech).

Skin friction—Parasite drag resulting from the contact of the air with the skin (and its imperfections) of the airplane.

Skiplane—Airplane equipped with skis for operation to and from snow or ice.

Skipper—A two-seat training aircraft manufactured by Beech Aircraft Corporation.

Sky cover—The amount of sky obscured by clouds or other phenomena. Generally described as clear, scattered, broken, overcast, obscured.

Skyhawk—A fixed-gear single-engine aircraft manufactured by Cessna Aircraft Company.

Skylane—A fixed or retractable gear high-performance single engine aircraft manufactured by Cessna Aircraft Company.

Cessna Skyhawk II (courtesy of Cessna).

Cessna Skylane RG (courtesy of Cessna).

Cessna Skymaster (courtesy of Cessna).

Skymaster—A family of push-pull twin-engine piston aircraft manufactured by Cessna Aircraft Company.

Skyrocket—A high-performance single-engine aircraft developed by Bellanca Aircraft Engineering (not to be confused with Bellanca Aircraft Corporation). At this writing it is not in production.

Skywagon—A fixed-gear single-engine six-place aircraft manufactured by Cessna Aircraft Company.

Bellanca Skyrocket II.

Cessna Skywagon (courtesy of Cessna).

Slant range—The line from flight altitude to a given point on the ground. For instance, the slant range from an aircraft positioned five miles horizontally from a VORTAC but 10,000 feet above the ground would be nearly seven miles. When used with reference to visibility, the slant range visibility frequently differs markedly from ground visibility.

Slash—A radar beacon reply represented on the scope by an elongated target.

SLD—Solid.

SLGT—Slight.

Sling load—A load carried by a helicopter underneath the aircraft.

Boeing Vertol civilian Chinook with sling load (courtesy of Boeing Vertol).

Slip—The tendency of an aircraft to lose altitude toward the center of a turn when not enough rudder or deliberate cross-control rudder is used for the degree of bank.

Slipstream—Propwash.

SLMM—Simultaneous compass locator at the middle marker.

SLO—Slow.

SLOM—Simultaneous compass locator at the outer marker.

Slow flight—Level flight just above stalling speed.

Slow roll—An aerobatic maneuver in which the aircraft, starting in level flight, rotates smoothly 360 degrees around its longitudinal axis and returns to level flight.

SLP—Slope.

SLR—Slush on runway.

SLS—Side lobe suppression.

SLT—Sleet.

sm—Statute mile.

Small aircraft—Light aircraft.

Smith (Ted) Aerostar Corporation—Developers of the Aerostar family of high-performance twins. Now a subsidiary of Piper Aircraft Corporation (Lock Haven, PA 17745.)

Smiths Industries, Inc.—Manufacturers of a variety of aircraft instruments. (P.O. Box 5389, St. Petersburg, FL 33518.)

SMK—Smoke.

SML—Small.

Smoke—Carbon particles suspended in the air resulting from fire. It tends to restrict flight as well as slant-range visibility. The sequence-report symbol is K.

SMTH—Smooth.

SMWHT—Somewhat.

SN—Snow grains (in sequence reports).

Snap roll—An aerobatic maneuver which is started at a speed approximately 30 knots above stalling speed. To induce a snap roll, bring the stick full back and simultaneously use aileron and full rudder in the

Piper Aerostar 601B (courtesy of Piper).

Mitsubishi Solitaire (courtesy of Mitsubishi).

desired direction of the roll. The nose comes up sharply and the roll takes place with great suddeness.

Snow—Precipitation in the form of ice crystals. In sequence reports the symbol is S.

SNRS—Sunrise.

SNST—Sunset.

SNW—Snow.

Soaring—Flying a glider or sailplane.

Soaring Society of America—(SSA)—See *aviation organizations*.

SOB—Souls on board.

Society of Automotive Engineers—(SAE)—See *aviation organizations*.

Socked in—Expression used when grounded by weather, especially fog.

Solar still—A primitive means of producing small amounts of drinkable water by using the heat from the sun. Dig a funnel-shaped hole in the ground, place a receptacle at its bottom, then cover the sides with plastic (if no plastic is available, most airplane seats are covered with non-porous material, which can be used for the purpose). The heat of the sun will cause moisture to be drawn from even the driest ground. It will collect on the underside of the plastic covering and drip into the receptacle.

Solitaire—One of the MU-2 family of corporate turboprop aircraft manufactured by Mitsubishi Aircraft International.

Solo—A flight during which the pilot is the only person in the aircraft. To solo usually refers to the first time the instructor leaves the aircraft and permits the student to take it up alone.

Sonic boom—Two loud bangs occurring in quick succession, produced by the leading and trailing edges of the wings of aircraft flying at supersonic speeds. Its severity depends on atmospheric conditions.

SOP—Standard operating procedure.

Sound barrier—A severe shock wave created by the airfoil as the aircraft passes from Mach .95 to Mach 1.05. It produces fluctuations in the

303

airspeed indicator for a few seconds and simultaneously rapid changes in the altimeter reading as the shock wave passes the static vents. As the speed increases, everything returns more or less to normal.

Sound, speed of—Mach 1. Approximately 600 mph under standard atmospheric conditions at sea level. See *Mach*.

Soup—Slang for clouds or fog surrounding the aircraft in flight. *Flying in the soup*.

SP—Snow pellets (in sequence reports).

Span—The distance from wing tip to wing tip.

Spar—The principal load-carrying beam running length-wise in an airfoil. In many aircraft the main spar also runs through the fuselage.

Sparker—An instrument carried aboard hot-air balloons for the purpose of igniting the pilot light on the burner.

Spark plug—Inserted into the combustion chamber, the spark plug uses electric current from a magneto to create an electric spark which ignites the fuel-and-air mixture. All piston engines certificated for use in aircraft have two spark plugs for each cylinder.

Spark-plug fouling—Deterioration of the ability of the spark plug to function properly, due to carbon deposits or burned tips, usually caused by operating with too rich or too lean a mixture.

Spats—Pants.

Speak slower—A pilot request to the controller to speak more slowly.

Special IFR—Aircraft which operate under a waiver or letter of agreement in control zones or terminal control areas. The pilots must be IFR rated and the aircraft IFR equipped.

Special use area—An area either on the surface or in the air or both, used for special (usually military) operations, such as jet training, missile firings. Flight through such areas may be prohibited, restricted, or pilots may be warned to be extra cautious.

Special VFR—(SVFR or S/VFR)—A rule which permits pilots to take off or land at controlled airports or fly through a control zone when conditions are below VFR minimums.

Special VFR conditions—Except at certain high-density airports where special VFR operations are not permitted, pilots may request special VFR clearances for takeoff, landing, or flight through the control zone if the visibility is at least one mile. There are no ceiling restrictions. The pilot must be in communication with the tower, follow the details of the clearance issued by the tower, but it is his responsibility to remain clear of clouds. Special VFR is never offered by ATC. It must be requested by the pilot and approved by ATC.

Special VFR operations—See *special VFR conditions*.

Speed adjustment—An ATC request to ask pilots to change the speed of their aircraft to a given higher or lower figure. It may be expressed as *increase speed to*(speed), or *reduce speed to* (speed), or *if feasible reduce (increase) speed to* (speed).

Speed brakes—Dive brakes.

Speed of sound—See *Mach*.

Sperry Flight Systems, Avionics Division—Manufacturers of high-performance avionics. (P.O.Box 29000, Phoenix, AZ 85023.)

Spin—An intentional or unintentional maneuver in which the airplane, after stalling, descends nearly vertically, nose low, with the tail revolving around the near-vertical axis of descent. Most modern light aircraft are placarded against intentional spins.

Spin recovery—Push the stick forward and apply rudder in the direction opposite to that of the spin. See *stall-spin accident*.

Spiral—A climb or descent in a constant turn of 360 degrees or more.

Split S—The maneuver following a half roll. It returns the aircraft from the inverted back to the normal attitude. Not to be attempted except at a safe altitude.

Spoilers—Flat vertical surfaces which can be raised out of the upper surface of the wing at the pilot's discretion. They spoil the airflow over the wing and cause it to lose lift and therefore drop. Spoilers may be constructed to be deployed simultaneously in both wings, or individually in one wing or the other. All gliders and sailplanes are equipped with simultaneously deploying spoilers to facilitate an increase in the rate of descent without an increase in speed, necessary in order to assure landing at the desired spot. Among powered aircraft, the MU-2 family of corporate turboprops uses individually deployable spoilers in the place of ailerons.

Sport—A two-place training aircraft manufactured by Beech Aircraft Corporation.

Spot landing—Accuracy landing.

SPRD—Spread.

SQAL—Squall.

SQLN—Squall line.

Squall—A strong local wind which starts and stops suddenly and lasts only a short time.

Squall—A severe local storm with attendant gustiness and intense precipitation.

Beechcraft Sport 150 (courtesy of Beech).

Squall line—A line or band of active connected thunderstorms, not associated with a front.

Squawk—Expression used to mean activating the transponder.

Squawk-ident—Activate the transponder and push the ident button. An ATC phrase.

Squelch—A means of adjusting a radio receiver to eliminate background noise and static from the selected frequency. Most of the more advanced modern com radios have an automatic squelch feature.

SR—Sunrise.

SS—Sunset.

SSA—Soaring Society of America.

SSALR—Simplified short approach light system with RAIL.

SSB—Single side band.

SST—Supersonic transport.

Stabilator—A horizontal tail surface which moves up or down in its entirety to control pitch movement of the aircraft. It is, in fact, a stabilizer-elevator combination.

Stability—The ability of an aircraft to remain at a constant heading, altitude or other condition without being affected by outside forces. A stable aircraft will return to its original attitude after a brief disturbance without control input by the pilot. No known aircraft will maintain straight and level flight indefinitely without assistance from the pilot or autopilot. Thus, 100 percent stability has not yet been achieved.

Stability augmentation system—An automatic autopilot-flight-control system in helicopters. Mandatory for IFR flight.

Stabilizer—The fixed horizontal or vertical tail surface which gives an aircraft stability.

Stable air—Air which tends to remain at a fixed altitude. It tends to hold pollution and other obstructions to visibility in place.

Stage I, II, III service—See *terminal radar*.

Stall—The inability of an airplane to continue flight due to an excessive angle of attack. It will either drop its nose and thus reduce the angle of attack and regain flying speed, or, if forced to retain the excessive angle of attack and control surfaces are moved, either intentionally or unintentionally, may fall into a spin.

Stalling angle of attack—Burble point.

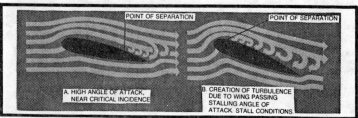

Stall (courtesy of FAA).

Stalling speed—The speed at a given angle of attack at which airflow separation begins and a stall occurs. Aircraft can stall at virtually any speed if an acceptable angle of attack is exceeded. In normal use the term means the slowest speed at which the aircraft can barely maintain straight and level flight. It varies with different types or aircraft and with flap and gear position.

Stall-spin accident—A common form of usually fatal accident in which the aircraft is stalled close to the ground and in an effort to avoid hitting the ground the pilot increases backpressure on the yoke and uses ailerons or rudder, causing the aircraft to spin at an altitude from which recovery is not possible.

Stall warning—A device, usually involving a buzzer, a light or both, which indicates to the pilot that the aircraft is about to stall. It is also the noticeable change in the feel of the controls which indicates that a stall is imminent.

Standard atmospheric conditions—See *standard conditions*.

Standard broadcast station—Any radio station. ADF-equipped aircraft can use standard broadcast stations for navigational guidance. Many such stations are shown on aeronautical charts.

Standard class—With reference to sailplanes, in U.S. competition soaring, a class of sailplane, the wing span of which may not exceed 15 meters (49 feet 2 inches).

Standard conditions—(ISA)—The abbreviation stands for ICAO Standard Atmosphere. An atmospheric pressure related to the temperature which serves as a basis for comparing actual conditions: A pressure of 29.92 in Hg at sea level with a temperature of 59 degrees F. and a lapse rate of about 3.5 degrees F. per 1,000 feet of altitude.

Standard instrument approach procedure—See *instrument approach procedures*.

Standard instrument departure—(SID)—A preplanned IFR departure procedure devised by ATC and printed for use by pilots. Use of SIDs reduces the length of IFR clearances and the workload for controllers. In order to accept a clearance involving a SID, the pilot must have the printed instruction with him in the cockpit.

Standard rate turn—A turn of three degrees per second requiring two minutes for a 360-degree turn.

Standard terminal arrival route—(STAR)—A preplanned IFR arrival procedure devised by ATC and printed for use by pilots. Use of STARs reduces controller workload. In order to accept a clearance involving a STAR, the pilot must have the printed instruction with him in the cockpit.

Stand by—In aviation-radio phraseology a request to wait on that frequency for further transmissions.

Stand by one—Same as stand by, suggesting that the wait will be short (*one* could mean second or minute or moment).

STAR—Standard terminal arrival route.

Standard Instrument Departure (SID)
MEETS FAA REQUIREMENTS FOR AERONAUTICAL CHARTS

SAN ANTONIO, TEXAS
SAN ANTONIO INTL

HENLY FIVE DEPARTURE
(HENLY5•HENLY) (PILOT NAV)
FOR TURBOJET AIRCRAFT ONLY

TAKE-OFF
Rwy 12R: Maintain runway heading until 10 DME from San Antonio VORTAC, then turn LEFT direct to San Antonio VORTAC. Cross San Antonio VORTAC at or above 9000'. Thence
Rwys 3R, 21L & 30L: Maintain runway heading for radar vector to San Antonio R-359 Thence

DEPARTURE
Via San Antonio R-359 to Henly Int, then via assigned route.
For rwy 12R, cross San Antonio R-359/8 DME at or above 11000'

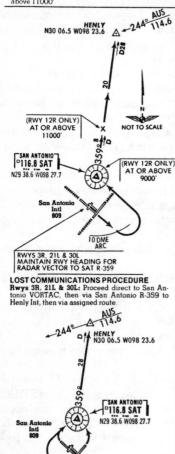

LOST COMMUNICATIONS PROCEDURE
Rwys 3R, 21L & 30L: Proceed direct to San Antonio VORTAC, then via San Antonio R-359 to Henly Int, then via assigned route.

LEJON ONE DEPARTURE
(LEJON1•LEJON)
FOR TURBOJET AIRCRAFT ONLY

TAKE-OFF
Rwy 12R: Maintain runway heading until 10 DME from San Antonio VORTAC, then turn LEFT direct San Antonio VORTAC, then via San Antonio R-301. Cross San Antonio VORTAC at or above 9000'. Cross San Antonio R-301/5 DME fix at or above 10000'. Thence
Rwys 3R, 21L & 30L: Maintain runway heading for radar vector to San Antonio R-301 Thence

DEPARTURE
Via San Antonio R-301 to Lejon Int, thence via (transition) or (assigned route).
TRANSITION
Sheffield (LEJON1•8FF): From Lejon Int to Sheffield Int (118 nm): Via 260° heading to intercept J-138.

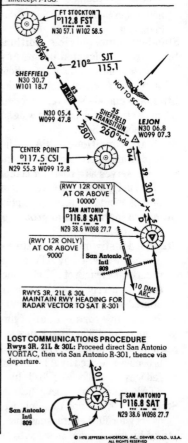

LOST COMMUNICATIONS PROCEDURE
Rwys 3R, 21L & 30L: Proceed direct San Antonio VORTAC, then via San Antonio R-301, thence via departure.

CHANGES: Lejon Int coords revised.

Standard instrument departure (SID) chart (courtesy of Jeppesen Sanderson, Inc.).

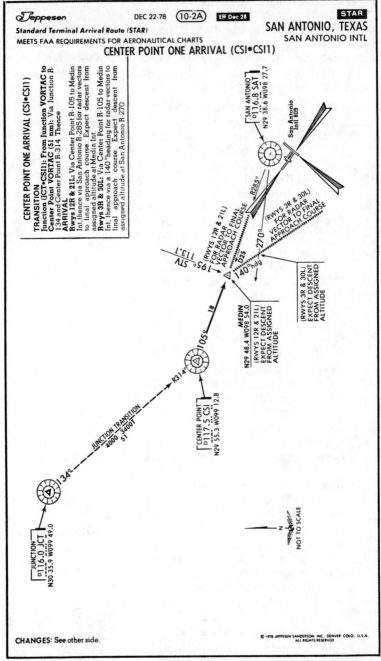

Standard terminal arrival route (STAR) chart (courtesy of Jeppesen Sanderson Inc.).

Starter—Ignition switch and the starter motor which must turn the engine in order to start combustion and ignition.

State aviation departments—The departments or agencies of various states, dealing with aviation-related matters. They may in no case countermand FAA or ATC regulations.

Alabama Department of Aeronautics
    11 South Union Street, Montgomery, AL 36104
Arizona Department of Aeronautics
    3000 Sky Harbor Blvd., Phoenix, AZ 85034
Arkansas Division of Aeronautics
    1515 Building, Little Rock, AR 72202
California Department of Aeronautics
    Executive Airport, Sacramento, CA 95822
Connecticut Department of Transportation
    24 Wolcott Hill Road, Wethersfield, CT 06109
Delaware Aeronautics Section
    P.O.Box 778, Dover, DE 19901
Florida Department of Transportation
    Tallahassee, FL
Georgia Department of Transportation
    2 Capital Square, Atlanta, GA 30334
Idaho Department of Aeronautics
    3103 Airport Way, Boise, ID 83705
Illinois Department of Aeronautics
    Capital Airport, Springfield, IL 62705
Indiana Aeronautics Commission
    100 North Senate Avenue, Indianapolis, IN 46204
Iowa Aeronautics Commission
    State House, Des Moines, IA 50319
Kansas Department of Economic Development, Aviation Division
    State Office Building, Topeka, KS 66612
Kentucky Department of Aeronautics
    Plaza Tower, Frankfort, KY 40601
Louisiana Dept. of Public Work, Aviation Division
    P.O.Box 44155, Capital Station, Baton Rouge, LA 70804
Maine Department of Aeronautics
    State Airport, Augusta, ME 04330
Maryland State Aviation Administration
    P.O. Box 8755, Baltimore, MD
Massachusetts Aeronautics Commission
    Boston Logan Airport, East Boston, MA 02128
Michigan Aeronautics Commission
    Capital City Airport, Lansing, MI 48906
Minnesota Department of Aeronautics
    Downtown Airport, St. Paul, MN 55107

Mississippi Aeronautics Commission
P.O.Box 5, Jackson, MS 39205
Missouri Division of Commerce and Industrial Development
Aviation Section, P.O.Box 118, Jefferson City, MO 65101
Montana Aeronautics Commission
P.O.Box 1698, Helena, MT 59601
Nebraska Department of Aeronautics
P.O.Box 82088, Lincoln, NE 68501
Nevada Public Service Commission
Carson City, NV
New Hampshire Aeronautics Commission
Municipal Airport, Concord, NH 03301
New Jersey Department of Transportation
1035 Parkway Avenue, Trenton, NJ 08625
New Mexico Department of Aviation
P.O.Box 579, Santa Fe, NM 87501
New York State Department of Transportation
1220 Washington Avenue, Albany, NY
North Carolina Department of National and Economic Resources
P.O.Box 27687, Raleigh, NC 27611
North Dakota Aeronautics Commission
Municipal Airport, Box U, Bismarck, ND 58501
Ohio Department of Commerce
3130 Case Road, Columbus, OH 43220
Oklahoma Aeronautics Commission
5900 Mosteller Drive, Oklahoma City, OK 73112
Oregon State Board of Aeronautics
Salem, OR
Pennsylvania Department of Transportation, Bur. of Aviation
Capital City Airport, New Cumberland, PA 17070
Rhode Island Department of Transportation
Green State Airport, Warwick, RI 02886
South Carolina Aeronautics Commission
P.O.Box 88, West Columbia, SC 29169
South Dakota Aeronautics Commission
Pierre, SD 57501
Tennessee Department of Transportation
P.O.Box 17326, Nashville, TN 37217
Texas Aeronautics Commission
P.O.Box 12607, Austin, TX 78711
Utah Division of Aeronautics
International Airport, Salt Lake City, UT 84116
Vermont Aeronautics Board
State House, Montpelier, VT 05602
Virginia State Corporation Commission, Div. of Aeronautics
P.O.Box 7716, Richmond, VA 23231

Washington State Aeronautics Commission
  8600 Perimeter Road, Seattle, WA 98108
West Virginia State Aeronautics Commission
  Kanawha Airport, Charleston, WV 25311
Wisconsin Department of Transportation
  951 Hill Farms Office Bldg., Madison, WI 53702
Wyoming Aeronautics Commission
  P.O.Box 2194, Cheyenne, WY 82001

Static—Radio-reception interference resulting in noise.

Static—Radio-reception interference resulting in noise.

Static pressure—Undisturbed atmospheric pressure as received by the static vents in an aircraft.

Static vent—A hole, usually located in the side of the fuselage, which provides air at atmospheric pressure to operate the pitot-static system. There are usually two such vents, one in each side, connected by a Y-shaped tube from which the pitot-static instruments obtain their reading. The dual installation prevents sideways movement from affecting the accuracy of the readout.

Stationair—A six or seven-seat fixed-gear single-engine aircraft manufactured by Cessna Aircraft Company.

Stationary front—A weather front which has stopped moving or at least is moving at less than five mph, or has matching temperatures on each side. The weather associated with stationary fronts tends to spread and remain for several days. The weather-chart identification is a line with pointed marks on one side and rounded marks on the other.

Stationary reservations—Altitude reservations related to military operations.

Station passage—The time at which an aircraft passes over a fix.

Station report—A group of symbols on a weather chart showing conditions at a specific weather-reporting station. Conditions reported include wind-force equal to Beaufort scale number, wind direction, cloud type

Cessna Stationair 7 (Cessna).

and altitude, sea-level pressure, barometric tendency and precipitation occurring since the previous report. Reports are issued every six hours.

Statute mile—(sm)—5,280 feet; .87 of a nautical mile.

STBL—Stable.

STC—Supplementary type certificate.

STDY—Steady.

Stearman—A high-performance single-engine biplane, often used in agricultural operations.

Steep turn—Generally any turn involving a bank in excess of 30 degrees.

STG—Strong.

Stick—Control wheel; yoke.

Stick pusher—A system built into certain jet aircraft which automatically forces the control wheel forward when the aircraft approaches a stall.

Stick shaker—A system built into certain jet and turboprop aircraft which shakes the control wheel quite violently in order to alert the pilot that he is on the edge of a stall.

Stiffeners—Members attached to the inside of the skin of an aircraft to help resist the effects of compression or bending loads.

STM—Storm.

STN—Station.

STOL—Short takeoff and landing (aircraft).

Stop altitude squawk—An ATC phrase to ask the pilot to turn off his altitude encoder.

Stop and go—A landing in which the aircraft comes to a full stop and then takes off again without returning first to the takeoff position on the runway.

Stopover flight plan—A single flight plan which includes one or more stops between departure airport and the final destination.

Stop squawk—An ATC phrase to ask the pilot to turn off his transponder.

Stopway—A continuation of the takeoff runway. It can support an aircraft during an aborted takeoff without causing structural damage. Also known as *overrun*.

Straight-and-level flight—Flight along a straight course without changes in altitude or airspeed.

Straight-in approach—A landing made without first flying a landing pattern. At controlled airports the tower will clear the pilot for a straight-in approach if traffic conditions permit and it reduces the time aloft for the pilot. At uncontrolled airports straight-in approaches are not advisable.

Straight-in approach IFR—An instrument approach in which the final approach segment is begun without first making a procedure turn. It does not necessarily have to culminate in a straight-in approach.

Straight-in landing—A landing made on a runway which is aligned within 30 degrees of the direction of the final approach course, at the completion of an instrument approach.

Strangle your parrot—Slang for *shut off your transponder*.

Stratiform—Clouds occurring in horizontal layers without much vertical development. They are the result of cooling layers in stable air and may occur in small or large patches or may cover the whole sky.

Stratocumulus—Low clouds with stratiform and cumuliform characteristics; most often forming below an inversion.

Stratosphere—The layer of the atmosphere above the troposphere and reaching to an altitude of about 30 miles. It is a region of fairly constant temperatures.

Stratus—Low stratiform cloud, usually found at the level of a temperature inversion.

Streamlining—Term used for eliminating the maximum number of causes of parasite drag.

Stress—Resistance of a body to an opposing force.

Stress—A consideration in the design of an aircraft, aimed toward construction which can withstand all loads exerted upon it.

Stress analysis—The science of analyzing the amount of stress resistance which must be built into an aircraft.

Strobe—A light which emits flashes of high intensity in rapid succession.

Strut—A rigid brace designed to support a given load. Wing struts, for instance, support the wing in terms of compression as well as tension loads.

Student pilot—A person receiving primary flight instruction.

Student pilot certificate—An authorization by the FAA, based upon recommendations by a flight instructor, permitting the student pilot to fly an aircraft solo within certain distance limitations. A student pilot certificate does expressly prohibit the student from carrying passengers. In order for a student pilot certificate to be effective, the student must have passed a third-class medical examination within the preceding 24 months.

S-turn—An S-shaped path flown on final approach in order to either lose altitude by increasing the distance, or to avoid overtaking a slower aircraft on final for landing.

S-turn—A coordination maneuver in which an S-shaped path is flown over reference points on the ground, allowing for wind drift.

Subsonic—Any speed below the speed of sound.

Substitute route—A route assigned to pilots on IFR flight plans when an airway or a portion thereof is unusable because of nav aid outages.

Sucker hole—Slang for something that looks like a safe clear path through a thunderstorm or a layer of clouds, but which can easily actually lead into areas of considerable turbulence or solid clouds.

Suction gauge—A cockpit instrument indicating the amount of suction produced by the vacuum pump. It should normally stay at about four in Hg to keep gyro instruments operating satisfactorily.

Sunair Electronics—Manufacturers of HF transceivers. (3101 SW 3rd Avenue, Ft. Lauderdale, FL 33315.)

Beechcraft Sundowner (courtesy of Beech).

Sundowner—A fixed-gear single-engine aircraft manufactured by Beech Aircraft Corporation.

Supercharger—Turbocharger.

Supercritical wing—A term used for an airfoil design created by NASA with the help of computers. It has low drag and high lift characteristics and results in a reduction in fuel consumption and an increase in range.

Super Cub—A fixed-gear single-engine aircraft manufactured by Piper Aircraft Corporation.

Supersonic—Speeds above the speed of sound (Mach 1 and up).

Super Viking—A family of high-performance single-engine aircraft manufactured by Bellanca Aircraft Corporation.

Supplementary type certificate—(STC)—A certificate issued by the FAA approving an addition to or modification of a certificated aircraft.

Surface forecast map—A weather chart of the earth's surface showing the positions that fronts, isobars, and pressure centers are expected to reach at a specific time.

Surface visibility—Ground visibility.

Surface wind—The wind blowing near the surface, of importance during approaches, takeoffs and landings. It is more likely to fluctuate than do winds aloft.

Surveillance radar—Terminal radar without the capability of determining altitude.

Piper Super Cub (courtesy of Piper).

315

Bellanca Super Viking (courtesy of Bellanca).

Surveillance radar approach—(ASR)—A ground-controlled instrument approach during which the controller gives vectoring information to the pilot and, during the final phase, reads off altitudes below which the pilot should not descend. It is a non-precision approach.

SVFR—Special VFR.

S/VFR—Special VFR.

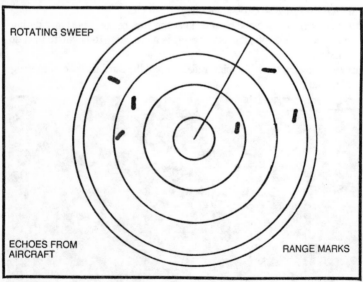

Surveillance radar scope. (ASR).

Swearingen Metro and Merlin(courtesy of Swearingen).

SVR—Severe.

SVRL—Several.

SW—Snow showers (in sequence reports).

SWAP—Severe weather avoidance plan.

Swash plate—The device in a helicopter which transmits control inputs to the rotor assembly.

Swearingen Aviation Corporation—Manufacturers of the Merlin and Metro families of turboprop corporate and air-taxi aircraft. (P.O.Box 32486, San Antonio, TX 78216.)

Swept-back wing—A wing on which both the leading and trailing edges are swept back from the lateral axis.

Swearingen Aviation Corporation turboprop aircraft specifications.

SWEARINGEN AVIATION CORPORATION MODEL:	MERLIN IIIB	MERLIN IVA	METRO II
ENGINES	GA/TPE 331-IOU-501G	GA/TPE-331-3U-304G	GA/TPE 331-3UW-304G
shp each	900	840	840 (dry) 940 (wet)
TBO (hours)	3,000	3,000	3,000 (6,000 commuter)
SEATS crew + passengers	2 + 6 to 9	2 + 10 to 13	2 + 20
TAKEOFF ground roll (ft)	NA	1,600 (1,350 short fld)	1,925 (1,350 short fld)
50' obstacle (ft)	3,219	2,620 (2,050 short fld)	2,620 (2,050 short fld)
RATE OF CLIMB (fpm)	2,782 (1 eng. 723)	2,400 (1 eng. 650)	2,400 (1 eng. 650)
Vxse (knots) (CAS)	119	126	126
Vyse (knots) (CAS)	138	133	133
Vmca (knots) (CAS)	107	91	91
ACCELERATE-STOP DISTANCE (ft)	NA	3,450 (3,200 short fld)	4,075 (3,100 short fld)
SERVICE CEILING (ft)	31,400 (1 eng. 16,500)	26,600 (1 eng. 14,700)	27,000 (1 eng. 14,700)
MAX SPEED (knots)	309 (12,000 ft)	269 (16,000 ft)	255 (19,000 ft)
BEST CRUISE SPEED (knots)	300 (15,000 ft)	263 (15,000 ft)	255 (10,000 ft)
LONG-RANGE CRUISE SPEED (knots)	272 (29,000)	244 (26,000 ft)	246 (21,000 ft)
RANGE w. res. long-range speed (nm)	2,278 (31,000 ft)	1,822 (25-28,000 ft)	2,139 (25,000 ft)
FUEL FLOW per engine (pph) best speed	370	313	349
long range speed	230	212	254
STALL clean/dirty (knots)	103/89	98/86	98/86
LANDING 50' obstacle (ft)	3,227	3,550 (1,970 short fld)	3,550
RAMP WEIGHT (lbs)	12,600	12,600	12,600
TAKEOFF/LANDING WEIGHT (lbs)	12,500/11,500	12,500/12,500	12,500/12,500
ZERO FUEL WEIGHT (lbs)	10,000	12,500	12,500
EMPTY WEIGHT (lbs)	7,800	8,200	NA
USEFUL LOAD (lbs)	4,700	4,300	4,300
FUEL CAPACITY (lbs)	4,342	3,712	4,342
MAX PAYLOAD (lbs)	2,000	4,100	4,300
PAYLOAD w. full fuel (lbs)	158	388	42
FUEL w. max payload (lbs)	2,500	NA	NA
WING LOADING (lbs per sq ft)	45	45	45
WING AREA (sq ft)	277.5	277.5	277.5
LENGTH/HEIGHT/SPAN external (ft)	42.16/16.82/46.25	59.34/16.7/46.25	59.34/16.76/46.25
LENGTH/HEIGHT/WIDTH cabin (ft)	17.4/4.75/5.17	33.1/4.75/5.17	33.1/4.5/5.17
PRESSURIZATION (psi)	7	7	7
PRICE (1979 $s)	1,216,000	1,275,000	1,275,000
STANDARD EQUIPMENT:			
DE-ICING WINGS, PROPS	Yes	Yes	Yes
PROPELLER SYNCHROPHASING	Yes	Yes	Yes
GALLEY	Yes	No	No
TOILET	Yes	Yes	No

Swinging the compass—Calibrating the compass, generally accomplished by comparing compass readings with a large compass rose painted permanently on the ramp at an airport. It must be done with the aircraft in flight attitude and with engines running and radios turned on.

Symolic display—Pictorial display.

Sympathetic resonance—A vibration which develops when the frequency of vibration in one mechanism is in phase with that of another. It tends to increase and can become destructive unless checked. It has been a serious problem in helicopters, but has been successfully eliminated by controlling the designs of gear boxes and related mechanisms.

Synoptic chart—Surface weather map.

SYS—System.

T—Takeoff (on instrument approach charts).

T—Tango (phonetic alphabet)

T—Thunderstorm (in sequence reports).

T—Tropical air mass.

T—True (after bearing).

Tab—Trim tab.

TAC—Tactical Air Command.

TACAN—Stands for tactical air navigation. A navigation system which was originally developed for the military. It gives distance and bearing information on UHF frequencies. It is slightly more accurate with reference to bearing information than a VOR and less subject to interference. Today, most TACANs are combined with VORs and called VORTACs, and any aircraft equipped with VOR plus DME or TACAN equipment can use the nav aids to receive distance and bearing information.

Tachometer—An engine instrument which displays the engine rpm (usually in 100s). It also includes a counter showing engine time based on average rpm and a colored line or bar shows the normal operating range(s).

Tactical air navigation—See TACAN.

TADS—Target acquisition and detection system.

Tail assembly—Empennage.

Taildragger—Popular phrase for tailwheel aircraft.

Tail-heavy—The tendency of an airplane to fly tail-low and nose-high which requires above-normal elevator control or nose-down trim. Usually caused by exceeding the rearward limits of the CG.

Tail-low—A flight attitude resulting from a tail-heavy condition of the aircraft. It is also the usual flight attitude during slow flight and during the flare just before touchdown.

Cessna Skywagon 185 taildragger (courtesy of Cessna).

Tailwheel gear—Landing gear with the third wheel under the tail and the main gear ahead of the CG. Also called *conventional gear*.

Tailwind—Wind blowing in the same direction as the line of flight.

Tailwind component—The amount of favorable effect on the ground speed resulting from wind blowing more or less in the direction of flight.

Takeoff—The phase of flight starting with the aircraft in position on the runway, but not yet moving, and ending at the moment of liftoff.

Takeoff distance—The length of runway needed to accelerate to liftoff speed. Ground run.

Takeoff-distance chart—Charts in aircraft flight manuals showing the takeoff distance under various conditions of airport elevation, wind, etc. Usually given for full-gross-weight conditions.

Takeoff horsepower—Maximum permissible horsepower.

Takeoff leg—Ground run.

Takeoff power—The brake horsepower developed under standard conditions with the engine running at full rpm and the manifold pressure at the maximum allowable setting.

Takeoff run—Ground run.

Takeoff speed—The airspeed necessary to develop sufficient lift to permit liftoff under prevailing density-altitude and aircraft-weight conditions.

Takeoff weight—The maximum permissible weight of the aircraft for takeoff. It usually coincides with maximum gross weight, but on certain larger aircraft it is slightly below the maximum weight permissible for taxi.

Tandem airplane—A two-seat aircraft with dual controls one behind the other.

Tandem seating—Single seats, one behind the other.

Tango—In aviation-radio phraseology the term used for the letter T.

Target—The blip on a radar scope resulting from a primary radar return.

Target symbol—A computer-generated symbol resulting from a primary radar beacon return.

TAS—True airspeed.

Task—In ballooning, any type of competition.

Taxi—To move the airplane on the ground or on water under its own power.

Taxi patterns—Patterns established by ATC to expedite traffic on the ground at busy airports.

Taxiway—Surfaces designed for the ground movement of aircraft. Their load-bearing capability is usually less than that of runways.

TBO—Time between overhauls. TBOs are established for engines and engine-related systems by the manufacturer, based on past experience with and performance of the engine and/or system. It is not an absolute guarantee that the engine will perform satisfactorily for the full TBO-hours. Conversely, exceeding the TBO, though not good operating practice, is not illegal.

TC—True course.

TCA—Terminal control area.

TCH—Threshold crossing height.

TDA—Today.

TDZE—Touchdown zone elevation.

TDZL—Touchdown zone lights.

Teardrop entry—One of three recommended (but not mandatory) means of entering a holding pattern. After crossing the holding fix the aircraft proceeds outbound on a track of 30 degrees or less to the holding course and then turns right, approximately 210 degrees, to intercept the holding course.

TEC—Tower-en-route control.

TECA—Tower-en-route-control area.

Technical Standard Order—(TSO)—A specific standard of quality established by the FAA with reference to avionics, instruments and equipment. TSOd products are usually somewhat more expensive than non-TSOd products.

Teledyne Continental Motors—Manufacturers of piston engines for aircraft. (Box 90, Mobile, AL 36601).

Telephone, airborne—See *airborne telephones*.

TEMP—Temperature.

Tension stress—Stress resulting from pull.

Terminal area—A loose term for the airspace in which approach-control service or terminal control service is provided.

Terminal area facility—The ATC facility providing terminal, approach- and departure-control services.

Terminal control area—(TCA)—Positive control airspace established around certain high-traffic airports. Shaped more or less like an inverted wedding cake, it extends in the center from the ground to a given altitude, and on the outer portions from a given floor to the same altitude. Within TCAs all aircraft, whether IFR or VFR, must maintain contact with ATC and operate in accordance with ATC instructions.

Terminal forecast—A weather forecast covering expected conditions at a given airport. It is prepared every six hours and refers to weather

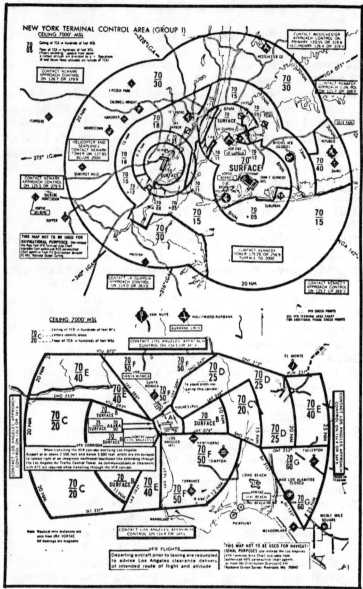

Terminal control area (TCA) charts for New York and Los Angeles (courtesy of FAA).

conditions expected within the following 12 hours. It uses sequence-report symbols and is issued for over 400 locations. In addition, 24-hour forecasts are available for about 130 major terminals.

Terminal instrument procedures—(TERPS)—Standard instrument procedures established for specific terminals, such as SIDs and STARs.

Terminal radar approach control—(TRACON)—A facility providing approach-control service for one or more airports using the same radar system.

Terminal radar program—A program in operation at certain airports which extends IFR-type separation service to VFR aircraft. Participation by VFR aircraft is optional. It is divided into Stage I: Traffic information and limited vectoring of VFR aircraft on a workload-permitting basis. Stage II: The same as Stage I, but on a full-time basis. Stage III: The same as Stage II, plus IFR-type separation between all participating aircraft. In view of the fact that VFR pilots may refuse to participate in the service, all aircraft in the area must maintain visual watch for other aircraft.

Terminal radar service area—(TRSA)—The airspace in which Stage I, II or III terminal radar service is provided.

Terminal VOR—(TVOR)—A low-power VOR co-located with an airport and used as a nav aid.

TERPS—Terminal instrument procedures.

Terrain clearance—The vertical distance between an aircraft in flight and the highest nearby point or object on the ground.

TET—Tetrahedron.

Tetrahedron—A rotating device on the ground, sometimes wedge shaped and sometimes in the shape of an airplane, showing the direction of the wind. At some locations tetrahedrons are equipped with flashing lights

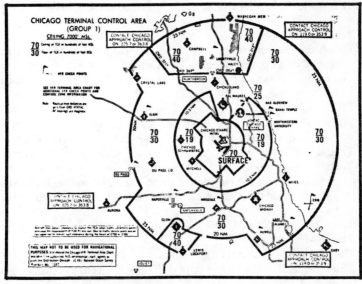

TCA chart for Chicago (courtesy of FAA).

which are turned on when the visibility is less than three miles. In addition, most are illuminated at night.

TFC—Traffic.

TH—True heading.

T-hangar—A long, narrow hangar building containing T-shaped spaces into which aircraft are pulled tail-first from either side. It is ideal for hangaring large numbers of light aircraft in a minimum amount of space.

That is correct—Aviation-radio phrase for *you have understood me correctly*.

THDR—Thunder.

Thermal—A prolonged updraft; especially important to glider pilots.

Thermaling—In soaring, using one or any number of thermals to gain and maintain altitude.

Thermal runaway—A phenomenon occurring with nickel-cadmium batteries, consisting of continuous and excessive increases in heat during charging. If left unchecked it can result in an explosion of the battery.

THK—Thick.

THN—Thin.

Three-point landing—A landing of a tailwheel aircraft in which the aircraft is stalled a few inches above the runway surface and all three wheels touch the ground at the same time.

Threshold—A line perpendicular to the centerline of the runway, indicating the beginning of the portion of the runway which is usable for landing.

Threshold crossing height—(TCH)—The height of the glide slope at the point where it crosses the threshold.

Threshold lights—A line of green lights across the runway at the threshold.

THRFTR—Thereafter.

Throttle—The knob or lever on the instrument panel or pedestal with which the pilot controls the quantity of fuel-air mixture fed to the engine.

Throttle—The valve which controls the amount of fuel-air mixture being fed to the engine.

THRU—Through.

Thrust—The forward force, pushing or pulling, exerted by the engine or, in the case of gliders, by gravity. It opposes and must overcome drag.

Thrust horsepower—The actual horsepower delivered by the engine.

Thrust reverser—A clamshell type of arrangement which, when deployed, reverses the direction of thrust from a jet engine. It reduces ground run and the need for excessive braking. Popularly called *tubs* or *buckets*.

THRUT—Throughout.

THSD—Thousand.

Thunderhead—A cumulonimbus cloud.

Thunder Pacific—U.S. Distributor of Thunder Balloons manufactured in England. (114 Sandalwood Court, Santa Rosa, CA 95401.)

Gulfstream-American Tiger (courtesy of Gulfstream-American).

Thunderstorm—A storm produced by cumulonimbus clouds accompanied by lightning, thunder, strong winds, turbulence, heavy rain and often hail. The vertical winds inside a thunderstorm are known to reach velocities of 100 mph and the associated turbulence is sufficient to tear an airplane apart.

Tiara—A new-design piston engine manufactured by Teledyne Continental Motors.

TIAS—True indicated airspeed.

Tiedown—Securing the aircraft with ropes or chains anchored to the ground in the tiedown area of an airport.

Tiger—A single-engine aircraft manufactured by Gulfstream-American Corporation.

TIL—Until.

Time group—A group of four digits representing the hours and minutes (0342 = 3:42 a.m.). Unless accompanied by a time-zone indicator, it is understood to represent Greenwich Mean Time.

Timers—See clocks.

Time zones—The various zones within which local time is different from one another; the time zones covering the U.S. are:

    Eastern Standard = GMT minus five hours
    Eastern Daylight = GMT minus four hours
    Central Standard = GMT minus six hours
    Cental Daylight = GMT minus five hours
    Mountain Standard = GMT minus seven hours
    Mountain Daylight = GMT minus six hours
    Pacific Standard = GMT minus eight hours
    Pacific Daylight = GMT minus seven hours
    Yukon Standard = GMT minus nine hours
    Alaska/Hawaii Standard = GMT minus 10 hours

Cessna Titan (courtesy of Cessna).

Tip stall—A stall occurring at the wingtip area. Tip stalls can result from turns at slow speed when the wingtip pointing to the center of the turn moves too slowly to generate lift.

TIT—Turbine inlet temperature.

Titan—A 10-seat turbocharged piston twin manufactured by Cessna Aircraft Company.

TKOF—Takeoff.

TMPRY—Temporary.

TMW—Tomorrow.

TNDCY—Tendency.

TNGT—Tonight.

To-from indicator—The portion of an OBI which shows whether the selected setting is a radial *from* the station or a bearing *to* the station. *Ambiguity meter*.

TOHP—Takeoff horsepower.

Tomahawk—A two-seat training aircraft manufactured by Piper Aircraft Corporation.

Torching—The burning of fuel in the exhaust outlet. It results from using an excessively rich mixture.

Tornado—A violently rotating column of air producing a funnel-shaped cloud which usually tips from a cumulonimbus cloud. It may touch ground leaving a path of destruction from a few feet to a mile wide. It tends to occur in the Midwestern states during late spring and early summer.

Torque—The normal tendency of an aircraft to rotate to the left in reaction to the right-hand rotation of the propeller. It alters with changes in power.

Total energy variometer—A variometer (vertical speed indicator used in gliders and sailplanes) which has been compensated so as to respond only to changes in the total energy of the sailplane.

Touch-and-go landing—A landing in which the aircraft does not come to a complete stop before starting another takeoff run. A popular but not terribly useful training exercise.

Touchdown—The point in the landing at which the gear of the aircraft first makes contact with ground or water.

Touchdown RVR—The runway visual range measured by equipment located near the touchdown zone of the runway.

Touchdown zone—The first 3,000 feet of the runway, beginning at the threshold.

Touchdown zone elevation—(TDZE)—The highest point in the touchdown zone; used to compute MDA or DH for instrument approaches.

TOVC—Top of the overcast.

Tower—Airport traffic control tower.

Tower controller—ATC controllers responsible for handling takeoffs and landings at a controlled airport.

Tower-en-route control—The control of IFR traffic between adjacent approach-control areas, eliminating the need to involve ARTCC. It expedites traffic handling and reduces the need for frequency changes.

Tower-en-route-control area—The area in which tower-en-route-control service is provided.

Tower frequency—The frequency or frequencies used by the control tower for air-to-ground communication. Also called *local control frequency*.

TPA—Traffic pattern altitude at uncontrolled airports.

TPG—Topping.

TPX 42—Numeric beacon decoder equipment providing rapid target identification, strengthening of the target return, and altitude information from aircraft equipped with Mode C capability.

Piper Tomahawk (courtesy of Piper).

Track—The imaginary line which the flight path of an airplane makes over the earth.

Tracking—Flying along a radio beam and, in the process, correcting for wind drift. Also following a certain ground feature, such as tracking a river.

TRACON—Terminal radar approach control.

Tracor, Inc.—Manufacturers of long-range navigation equipment. (6500 Tracor Lane, Austin, TX 78721.)

Trade winds—The general circulation pattern of easterly winds lying in two belts around the globe and divided by the doldrums. Extending to about 30 degrees north and south.

Traffic—Aircraft in flight.

Traffic advisories—Advisories given by ATC to IFR and often VFR aircraft with reference to other aircraft in such proximity or flying along a path which might result in a conflict.

Traffic information—Traffic advisories.

Traffic pattern—The pattern, consisting of downwind, base leg and final approach, which should always be flown by aircraft intending to land at an uncontrolled airport; also at controlled airports unless otherwise cleared by ATC.

Traffic in sight—Phrase used by pilots when sighting an aircraft mentioned by ATC in a traffic advisory.

Trailing edge—The usually sharp edge at the rear of any airfoil.

Trailing vortex—The vortex formed by air moving off the trailing edge of the wings.

TRANS—Transcribed.

Transceiver—A combination radio capable of transmitting and receiving.

Transcribed weather broadcast—(TWEB)—A continuous broadcast of transcribed meteorological information, available on LF/MF and some VOR facilities.

Transfer of control—Handoff.

Transferring controller—An ATC controller transferring control of a given aircraft to another controller.

Transferring facility—An ATC facility transferring control of a given aircraft to another facility when that aircraft leaves the air space under its jurisdiction.

Transition—A change from one phase of flight to another (climbout to cruise). Also SIDs and STARs which are designed as standard transitions between the en-route portion of the flight and the approach or takeoff phases.

Transition area—The airspace, usually between 700 and 1,200 feet agl, used in conjunction with instrument approach procedures at an airport.

Transition zone—The region of changing weather conditions occupied by a front. It can vary greatly in width.

Translation lift—In helicopter flight the additional lift which develops with forward speed and increases with increased forward speed.

Transponder (courtesy of Collins).

Transmissometer—A device used to measure horizontal visibility.

Transmitter—The portion of a com radio which transmits. Usually combined with a receiver in one unit.

Transmitting (in the) blind—Transmitting under circumstances where two-way communication cannot be or has not been established, but where the person transmitting believes that his transmission is being received.

Transponder—An airborne radar-beacon transceiver that automatically transmits responses to interrogation by ground-based transmitters.

Traveler—A single-engine aircraft manufacturered by Grumman-American Corporation.

Triangulation—A method of plotting the position of an aircraft at the intersection of two bearing lines to two known points, usually two VORs. While it can be accomplished with one nav receiver, it is easier and more accurate with two.

Tricycle landing gear—A three-wheel landing gear with the two main wheels located aft of the CG and the third wheel under the nose of the airplane.

Triggering time and temperature—In soaring, the time and temperature at which usable thermals begins to form.

Trim—Adjustment of the pitch attitude of an aircraft to achieve the desired attitude without added elevator input.

Trim tab—A small airfoil attached to a control surface, primarily the elevator, which makes minor adjustments in the position of that control surface under varying flight conditions. It may be fixed or hinged. Hinged trim tabs can be adjusted by the pilot in flight. Fixed tabs are installed to correct permanent stability problems in the airplane design.

TRML—Terminal.

TRNG—Training.

TROF—Trough.

Tropical air mass—A warm air mass originating in the tropical regions. It may be continental or maritime.

Tropopause—The region at the top of the troposphere.

Troposphere—The portion of the atmosphere which is below the stratosphere. It extends outward seven to 10 miles from the earth's surface and

Transponders comparison chart.

TRANSPONDERS

MANUFACTURER	MODEL	PRICE uninstalled	Volts input DC	Volts input AC 400 Hz	Output, watts	Units	Weight lbs.	REMARKS
RADAIR	250	595	14/28		250	2	3.2	
NARCO	AT 150	625	14/28*		250	1	2.5	*28V ADAPTER S145
GENAVE	BETA-5000	625	14/28*		250	1	3.4	*28V ADPATER S65
COLLINS	TDR-950L	645	14		150	1	2	
KING	KT 78A	655	14/28*		150	2	3.7	*28V ADAPTER S10
KING	KT 76A	695	14/28*		250	2	3.5	*28V ADAPTER S10
EDO-AIRE/FAIRFIELD	RT-777	695	14/28*		500	2	3.4	*29V ADAP1EH S60
COLLINS	TDR-950	695	14		250	1	2	
ARC	300/ RT-359A	700	28		125	1	3	
EDO-AIRE/FAIRFIELD	RT-668	795	14/28*		500	2	4	*28V ADAPTER S60
ARC	400/ RT-459A	800	28		250	1	3	
BENDIX AVIONICS	TPR-2060	820	14/28*		250	1	2.9	*28V ADAPTER S23
EDO-AIRE/FAIRFIELD	RT-887	945	14/28*		500	2	4	*28V ADAPTER S60
ARC	800/ RT-859A	1,300	28		250	1	3	
NARCO	AT 200	1,450	14/28*		250	1	2.8	*28V ADAPTER S145. PROGRAMMABLE AND PRESET CODE STORAGE AND RECALL. ELECTRONIC DIGITAL DISPLAY
COLLINS	TDR-90	2,345	28		500	1	3.5	
KING	KXP 755	2,515	28		700	3	6.1	
KING	KXP 7500	5,640.50	28		700	1	9	DOES NOT INCLUDE CONTROL HEAD. REPLY LIGHT PER ARINC CHARACTERISTIC 572
COLLINS	621A-6A	5,669		115	750	1	13	SELF TEST AND REPLY. FAULT ANNUNCIATION
BENDIX AVIONICS	TRA-63A	7,428		115	500	2	12	

Additional columns (shaded indicator matrix): Modes (A, B, C, D); Reply light (automatic, manual, dimmer); Remote ident avail.

Traveler.

within it temperature decreases rapidly with altitude, clouds form and convection is active. It is the portion of the atmosphere in which most aircraft operate.

Trough—An elongated area of low atmospheric pressure with the lowest pressure at the centerline of the trough.

TRRN—Terrain.

TRSA—Terminal radar service area.

True air speed—(TAS)—The actual airspeed of an aircraft relative to the air through which it is moving. It is calibrated airspeed adjusted for actual air density and altitude. It can be calculated from indicated airspeed by using the correction scale of a flight computer.

True altitude—Actual height above sea level, expressed in feet msl.

True course—(TC)—The intended flight path expressed as the angle in degrees between a meridian near the middle of the flight path, and the flight path itself.

True heading—(TH)—The true course with a wind-correction angle added or subtracted as needed.

True indicated airspeed—Calibrated airspeed.

True north—The geographic (not magnetic) north. The direction of the north pole from any point on the globe.

TSHWR—Thundershower.

TSMT—Transmit.

TSMTG—Transmitting.

TSMTR—Transmitter.

TSO—Technical standard order.

TSO—Time since overhaul.

TSTM—Thunderstorm.

Tubs—Slang for thrust reversers.

Tumble—To fall out of operating position, such as a gyro or compass. Most gyro compasses will tumble after a 55-degree bank and must then be reset with the caging mechanism.

TURBC—Turbulence.

Turbine—Jet engine.

Turbocharger—A turbine driven by the exhaust gases which compresses air and thus increases the amount of fuel-air mixture available to the engine. Turbocharged light aircraft usually use either of two types of installations. One type employs a second throttle to activate the turbocharger when the aircraft has reached an altitude at which power begins to decrease because of decreasing atmospheric pressure. The other type of installation is fully automatic and requires no special attention from the pilot.

Turbocharging—Increasing the pressure of the air used in the fuel-air mixture to exceed the available atmospheric pressure available at a given altitude.

Turbofan—A high-bypass jet engine equipped with a multi-blade fan which acts more or less like a shrouded propeller and delivers a considerable percentage of the total thrust. Fanjets are quieter and more economical in terms of fuel flow than pure jets, but they lose some of their effectiveness at altitudes above 35,000 feet or so.

Turbojet—A pure jet engine.

Turboprop—An aircraft powered by jet engines which drive propellers More economical to operate than pure jets or fanjets, they are somewhat slower and the best operating altitudes are usually below 30,000 feet.

Turboshaft—A jet-engine installation in a helicopter in which the engine drives the shaft which supports the rotorblades.

Turbulence—A disturbance or irregularity in the movement of the air. It can be caused by friction when the air moves over an uneven surface, or by the mixing of several currents of air with differing velocities. Turbulence is described as light, moderate, severe, or extreme. Light turbulence is uncomfortable but has no serious effect on the pilot, or on objects carried in the airplane. Moderate turbulence may cause unsecured objects to tumble or even fly around in the cockpit. Severe turbulence results in considerable fluctuations in airspeed and occupants of the aircraft are forced violently against their seatbelts The airplane may occasionally seem to be out of control as roll and pitch movement are briefly greater than can be counteracted by the controls. Extreme turbulence means that the airplane is virtually uncontrollable and is simply being tossed about. Since different aircraft and different pilots react differently to the varying degrees of turbulence, turbulence reports tend to be unreliable. Such reports always include the type of aircraft from which the report originated, but a 5,000-hour pilot flying a heavy twin may report a turbulent condition as light while a 200-hour pilot flying a Skyhawk may call the same condition moderate or even severe.

Turn-and-bank indicator—A gyroscopic flight instrument which displays the rate of turn and shows whether the turn is properly coordinated. It usually consists of a needle showing the rate of turn and a ball which will remain centered in a slightly curved glass tube, if the turn is properly coordinated. If the ball moves to either side, it indicates the need for increased rudder pressure on the side to which the ball is moving.

Turnaround—A semicircular taxiway adjacent to the ends of a runway, usually found at small airports where the runway is used for taxiing as

Aerospatiale TwinStar (courtesy of Aerospatiale).

well as takeoffs and landings. It permits the airplane to turn around or to get out of the way of other traffic using the runway.

Turn coordinator—Turn-and-bank indicator.

Turning radius—An aircraft making the tightest possible self-powered turn either on the ground or in the air describes a circle. The turning radius is, in fact, the distance from the center of that circle to the circle itself. While this is technically correct, the diameter of the circle is often described (incorrectly) as the turning radius.

TV—Television.

TVOR—Terminal VOR.

TWD—Toward.

TWEB—Transcribed weather broadcast.

Twenty-four-hour clock (time)—The time used in aviation-radio communication (as well as by the military and by several European countries). It consists of four digits (time group), the first two representing the hours from 01 to 24, and the second two representing the minutes from 00 to 59. In spoken communication *zero four five one* would indicate 4:51 a.m. and *two one zero six* would indicate 9:06 p.m.

Twin-engine aircraft—An aircraft powered by two engines. Usually used with relation to piston engines. Aircraft powered by turbine engines are referred to as twin-jet, twin-turboprop, or twin-turbine helicopter.

TwinStar—A twin-turbine helicopter manufactured by Aerospatiale in France.

Two-minute turn—A standard-rate turn, requiring two minutes to complete 360 degrees of turn.

Two-way radio communication failure—A situation for which, if it occurs in IFR flight, certain specific procedures have been developed by ATC (FAR Part 91.127). The transponder code to alert ATC to the fact that communication capability has been lost is 7600.

TWR—Tower.

TXWY—Taxiway.

Type certificate—The certificate issued by the FAA for a particular type of aircraft, certifying its airworthiness and stating operational limitations.

Typhoon—A Pacific hurricane or tropical cyclone.

333

U—Intensity unknown (in sequence reports).

U—Uniform (phonetic alphabet.)

U-1—Unicom at uncontrolled airport.

U-2—Unicom at controlled airport.

UDF—UHF direction finder.

UFN—Until further notice.

UFO—Unidentified Flying Object.

UHF—Ultra high frequency.

Ultimate load—The load required to cause a structure to fail.

Ultra high frequency—(UHF)—The band of electromagnetic frequencies from 300 to 3,000 MHz.

Ultra-light aircraft—A lightweight aircraft having a wing loading of less than three pounds per square foot (such as hang gliders).

Unable—Aviation-radio term stating that the pilot is unable to comply with an instruction or clearance or that a controller cannot approve a route or altitude request.

UNAVBL—Unavailable.

UNCLTD—Uncontrolled.

Uncontrolled airport—An airport without an operating control tower. Airports with part-time control towers are uncontrolled during the hours when the tower is not in operation.

Uncontrolled airspace—Airspace in which an appropriately rated pilot may operate IFR without being in contact with ATC. Airspace over which ATC has no jurisdiction.

Undershoot—To touch down short of the intended point of landing.

Under the hood—The pilot is wearing a hood which restricts his visibility to the instrument panel. Used during flight instruction for an instrument rating. An appropriately rated pilot must occupy the other seat and maintain a watch for other traffic.

Unicom—The radio frequencies assigned to aeronautical advisory stations for communication with aircraft. Unicoms are usually manned by FBOs or airport personnel and provide pilots with such information as the active runway, wind direction and velocity and other conditions of importance to the pilot. Unicoms are not authorized to give takeoff, landing or other clearances, though they may, at times, be utilized by ATC to relay clearance, in which case the transmission must be preceded by *ATC clears…* The unicom frequencies are 122.8 and 123.0 MHz at uncontrolled airports and 122.95 MHz at controlled airports.

Uniform—In aviation-radio phraseology the term used for the letter U.

United Instruments—Manufacturers of encoding altimeters. (2415 South Glendale, Wichita, KS 67201.)

United States—The 50 states, District of Columbia, Puerto Rico, all possessions including territorial waters and the airspace above.

United States Parachute Association—See *aviation organizations*.

UNLGTD—Unlighted.

Unlimited ceiling—A sky-cover report indicating that the sky is clear or has scattered clouds, as observed from the ground. The sequence report symbol is W.

Unlimited class—In U.S. competition soaring, a class of aircraft without specific requirements or restrictions. In air racing any piston-engine aircraft without limit to the horsepower.

Unlimited visibility—Horizontal visibility in excess of 15 miles.

UNMRKD—Unmarked.

Unpublished route—A route for which no minimum altitude has been established. It may be a direct route between two nav aids, a radar vector, a radial from a nav aid or portions of an approach beyond the limits of the published instrument approach procedure.

UNRSTD—Unrestricted.

Unstable air—Turbulent or gusty air which tends to move easily and continues to do so when displaced.

UNSTBL—Unstable.

Unusable fuel—Fuel in the tanks of aircraft which may not be able to reach the fuel lines at all or only when the aircraft is in steady straight and level flight. It is subtracted when calculating the range available with the amount of fuel on board. It is included in the empty weight of an aircraft.

Unusual attitudes—Any attitude in terms of pitch or roll which is beyond the normal operating attitude. Recovery from unusual attitudes by reference to instruments alone is an integral part of flight training for an instrument rating.

Updraft—A convection current moving upward. A thermal.

Upper winds—Winds aloft.

UPR—Upper.

UPSLP—Upslope.

Upwind—Into the wind.

Upwind leg—The flight path parallel to the landing runway and in the same direction as the intended landing.

Upwind side—The side of a mountain or other terrain or man-made feature toward which the wind is blowing.

URD—Utterance recognition device.

US—United States.

Unusable fuel—The portion of fuel in the tanks which can be drawn into the fuel lines in all usual flight attitudes. It is included in the useful load figure.

USAF—United States Air Force.

USB—Upper side band.

USCG—United States Coast Guard.

Useful load—The weight which can be carried in an aircraft in addition to the empty weight and without exceeding the maximum gross weight. It differs from *payload* in that *useful load* includes useable fuel and the professional crew and its baggage (in aircraft flown by a professional crew), while the *payload* includes neither.

USMC—United States Marine Corps.

USN—United States Navy.

Utility category aircraft—Light aircraft that will safely withstand 4.4 times its design gross weight in specific flying conditions, while aircraft licensed in the Normal Category require only 3.8 load limit factor.

UVDF—Direction finder operating in UHF and VHF frequencies.

V—Variable (in sequence reports).

V—Variation.

V—Victor (phonetic alphabet).

V—Symbol for speed. See *V-speeds*.

Vapor lock—Vaporization of fuel in the fuel lines which blocks the flow of fuel to the carburetor or injection system.

Vapor trail—Condensation trail.

VAR—Variation.

Variable-pitch propeller—A propeller the blade angle of which can be adjusted in flight or on the ground.

Variation—(V or VAR)—The angle between true north and magnetic north. It varies at different points on the globe because of local magnetic disturbances. It is shown on aeronautical charts as isogonic lines in degrees East or West and must be subtracted (east) or added (west) to the true course to get the magnetic course.

Variometer—A sensitive rapidly responding instrument showing the rate of climb or descent. A type of vertical-speed indicator used primarily in gliders and sailplanes.

VASI—Visual approach slope indicator.

VCNTY—Vicinity.

VDF—Direction finder using VHF frequencies.

Vector—A heading given to a pilot by a ground-based ATC radar facility.

Vector—The product of a combination of speed and direction or other forces.

Veering—Wind shifting in a clockwise direction, to the right of the direction from which it is blowing before. The opposite of backing.

Velocity—Speed (in a given direction). Most frequently used in relation to wind speeds.

Vertical development of cumulus clouds.

Ventral fin—A fin-shaped vertical stabilizer located either on top or at the bottom of the fuselage.

Venturi—A tube which is narrower in the middle than at either end. When air is forced through a venturi, it results in suction which can be used to drive gyro instruments on aircraft without vacuum pumps.

Venturi effect—The increase in wind velocity when wind blows down a valley or through a mountain pass.

Verify—In radio communication a request to check the information against the original source of that information to make sure it is correct.

Verify direction of takeoff (of flight after takeoff)—An ATC request used when an aircraft makes an instrument departure from an uncontrolled airport. It is usually relayed through a FSS or the unicom on the airport.

Vertical axis—The imaginary line running vertically through the fuselage at the CG. Also called *yaw axis*.

Vertical development—Cumulus clouds in the stage of building.

Vertical instruments—Engine and other instruments with a narrow vertical instead of a round display.

Vertical roll—An aileron roll executed while in vertical climb or descent.

Vertical separation—Separation of IFR traffic by altitude.

Vertical-speed indicator—(VSI)—An instrument, part of the pitot-static system, which indicates the rate of climb or descent in terms of fpm. It is usually calibrated in units of 100 fpm.

Vertical stabilizer—The fixed vertical airfoil on the empennage to which the rudder is attached.

338

Vertical speed indicator (courtesy of Smiths Industries).

Vertical stall—Whip stall or hammerhead stall. In a whip stall, the aircraft, having lost flying speed in a near-vertical attitude, slides briefly backwards, then pitches violently forward and down. In a hammerhead stall the pilot applies full rudder at the moment the airplane has lost upward momentum. The airplane turns sharply to the right (or left) and continues nose down.

Vertical takeoff and landing aircraft—Aircraft capable of liftoff without prior takeoff run. Primarily, helicopters.

Very high frequency—(VHF)—A band of electromagnetic frequencies between 30 MHz and 300 MHz.

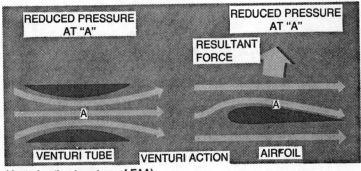

Venturi action (courtesy of FAA).

Very high frequency omnidirectional radio range—VOR.

Very high frequency omnirange—VOR.

Very high frequency omni test—VOR test facility.

Very low frequency—(VFL)—Electromagnetic frequencies below 30 kHz.

Vestibular sense—The function of the inner ear which provides a sense of balance. Unreliable under IFR conditions.

VFR—Visual flight rules.

VFR advisory service—Radar service available to VFR aircraft by numerous approach-control facilities when the aircraft intends to land at the airport served by the facility.

VFR aircraft—An aircraft operating VFR, or an aircraft not equipped to operate IFR.

VFR conditions—Weather conditions at or above the minimums required by the FARs for VFR operations.

VFR conditions on top—An IFR en-route clearance which allows the pilot to fly his aircraft at any appropriate VFR altitude above MEA which is at least 1,000 feet above cloud tops or obscuration.

VFR corridor—A corridor through some TCAs designed to permit VFR traffic to transmit the TCA without contacting ATC.

VFR flight—Flight conducted under visual flight rules in VFR weather conditions.

VFR flight plan—Filing a VFR flight plan assures the pilot that search and rescue operations will be activated automatically if he fails to reach his destination and the flight plan is not closed within 30 minutes of his ETA.

VFR low altitude training routes—Routes flown by military aircraft at or below 1,500 feet agl and at speeds exceeding 250 knots. They are flown only when ceilings are at least 3,000 feet and visibility is better than five miles. See *Olive Branch Routes, Oil Burner Routes*.

VFR on top—VFR flight conducted in VFR conditions above the tops of the overcast.

VFR over the top—VFR on top.

VFR tower—A control tower not equipped to provide approach-control services.

VHF—Very high frequency.

VHF omnidirectional radio range—VOR.

Victor—In aviation-radio phraseology the term used for the letter V.

Victor airway—A low-altitude airway between VORs. Victor airways are designated by the letter V followed by a number on the radio-facility charts.

Video map—An electronically displayed map, showing a wide variety of information of importance to the ATC controller. It can be called up on the radar scope by the controller.

Viking—A family of high-performance single-engine aircraft manufactured by Bellanca Aircraft Corporation.

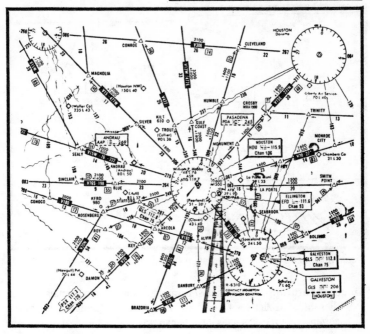

Victor airways (courtesy of FAA).

VIP—Very important person.

Visibility—The greatest horizontal distance at which an observer can identify prominent objects with the naked eye. Always reported in statute miles except when referring to RVR, in which case it is reported in feet.

Visual approach—An approach by an aircraft on an IFR flight plan, made by visual reference to the ground. It may be flown only if the conditions are VFR, and it must be authorized by the ATC.

Visual approach slope indicator—(VASI)—A two-color light system giving a clear indication to the pilot whether he is on, above or below the proper glide slope. When above the glide slope he sees only white lights. When on the glide slope he sees red light on the top and white light on the bottom. When below the glide slope he sees only red lights.

Visual flight rules—Minimum ceiling and visibility standards established by the FAA under which an aircraft may be operated by visual reference to the ground; plus all rules and regulations affecting VFR operation of an aircraft.

Visual holding—Holding an aircraft at an easily recognized geographical fix.

Visual meteorological conditions—Established minimums in terms of visibility, ceiling and distance from clouds.

341

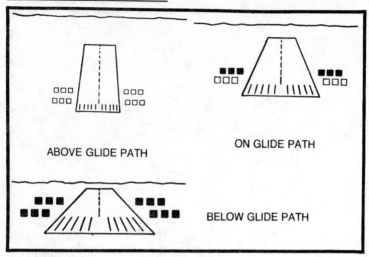

Visual approach slope indicator system (VASI) (courtesy of FAA).

Visual separation—A means of separating IFR aircraft in terminal areas. Either the controller sees the aircraft in question and issues instructions accordingly, or the pilot sees the other aircraft and, when cleared to do so by ATC, maintains his own separation.

VLF—Very low frequency.

VLF/Omega—A system of long-range (world-wide) area navigation, using a limited number of VLF transmitters to obtain position information.

VLY—Valley.

VNAV—Vertical navigation. A type of area-navigation equipment which includes a means of flying a controlled descent at a desired rate where no ground-based glide-slope information is available.

VOR—Very high frequency omnidirectional radio range. A nav aid consisting of two transmitters, one of which transmits a constant phase signal through 360 degrees of azimuth while the other rotates at 1,800 rpm and transmits a signal that varies with the constant-phase signal at a constant rate throughout the 360 degrees. This results in an infinite number of radials from the station (or bearings to the station) which are received by the nav receiver in the aircraft and displayed on the OBI. VOR frequencies range from 108.0 MHz to 117.95 MHz (not including the ILS localizer frequencies which are the odd-tenth decimal frequencies between 108.1 and 111.9 MHz). VORs are also referred to as *omni stations*.

VOR approach—A non-precision approach using one or several VORs as navigational reference.

VOR/DME—A combination of VOR and DME in one ground station.

VOR station—Usually a round flat-roofed structure with a cone-shaped antenna sticking up in the center. Always painted white.

VOR test facility—(VOT)—A ground transmitter that is designed to check the accuracy and proper functioning of the on-board VOR receiver while the aircraft is on the ground.

VORTAC—A combination VOR and TACAN at the same location, transmitting VOR and DME information.

Vortex—A rotating air mass created by the movement of an airfoil through the air; especially the trailing vortices which form aft of each wing tip from the tendency of high-pressure air below the wing to flow into the low-pressure area above the wing. See also *wake turbulence*.

VOT—VOR testing facility.

VR—Veering.

VRBL—Variable.

VSBY—Visibility.

VSI—Vertical-speed indicator.

V-speeds—Designations for certain speeds:

V_a = design maneuvering speed

V_b = design speed for maximum gust intensity

V_c = design cruising speed

V_d = design diving speed

V_{df} = demonstrated flight diving speed (M_{df} same in Mach)

V_f = design flap speed

V_{fc} = maximum speed for stability characteristics (M_{fc} = same in Mach)

V_{fe} = maximum flap-extended speed

V_h = maximum speed in level flight with maximum continuous power

V_{le} = maximum landing-gear-extended speed

V_{lo} = maximum landing-gear-operating speed

V_{lof} = lift-off speed

V_{mc} = mimimum control speed with the critical engine inoperative

V_{me} = maximum endurance speed

V_{mo} = maximum operating speed limit (M_{mo} = same in Mach)

V_{mu} = minimum unstick speed

V_{ne} = never-exceed speed

V_r = rotation speed

V_s = stalling speed or the minimum steady flight speed at which the aircraft is controllable

V_{so} = stalling speed or minimum steady flight speeding the landing configuration

V_{xse} = best single-engine angle-of-climb speed

V_y = best rate-of-climb speed

V_{yse} = best single-engine rate-of-climb speed

V_1 = critical engine-failure speed

V_2 = takeoff safety speed

V_{2min} = minimum takeoff speed

VTOL—Vertical takeoff and landing (aircraft)

W—Indefinite ceiling (in sequence reports).

w—Warm (description of an air mass).

W—West.

W—Whiskey (phonetic alphabet).

WAC—World aeronautical chart.

Wake turbulence—Turbulence created by the movement of an aircraft through the air. Primarily, the trailing wing-tip vortices which develop in the wake of heavy aircraft. Resembling a pair of counter-rotating horizontal tornadoes, they are at their worst behind a slow-moving heavy aircraft just after takeoff and just prior to touchdown. They tend to remain violent for several minutes, capable of throwing a smaller aircraft out of control. To avoid being caught in wake turbulence, liftoff should be accomplished before reaching the point of liftoff of the heavy aircraft, and should be followed by a steeper climb angle, preferably on the upwind side of the climb path of the heavy aircraft. During landing the light aircraft should stay above the flight path of the heavy aircraft (and, if possible, on the upwind side), and should touch down past the point at which the heavy aircraft touched down. The generation of wake turbulence ceases at the moment of touchdown of the heavy aircraft.

Walkaround—Preflight inspection.

Warm air mass—A mass of stable air warmer than the surface over which it is moving. It tends to cool as it moves, becoming even more stable, but reducing visibility.

Warm front—A usually deep front formed by a mass of warm, low-pressure air which is replacing a cold air mass. It is characterized by a sloping bank of clouds, steady precipitation, low ceiling and visibility and, in the winter, danger of icing in clouds. It is shown on weather charts as a line with rounded marks pointing in the direction of movement.

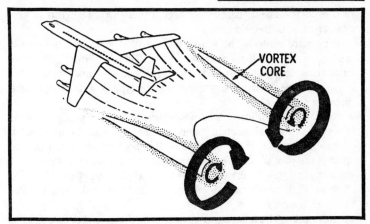

Wake turbulence pattern behind a heavy jet aircraft operating at low speed (courtesy of FAA).

Warmup—Letting the engine run at idle power until oil temperature and pressure are shown to be in the green.

Warning area—A special-use area off the coast of the U.S. and over international waters. Identified on charts by a W.

Washout—A decrease in the angle of incidence built into the tip areas of the wings to cause wing tips to fly at a lower angle of attack than wing roots. This prevents the tips from stalling during slow-flight maneuvers, and assures effective aileron control at low speeds.

Water vapor—Water in gaseous form in the atmosphere, primarily below 30,000 feet. When condensing it produces rain, snow, fog, dew, etc.

WAVE—Wind, altimeter voice equipment.

Wave-off—Being told not to land by someone stationed on the ground using visual signals. Primarily in the context of landing on an aircraft carrier.

Waypoint—(WP)—A navigational fix used in area navigation, created by electronically relocating a VORTAC from its actual position to a position desired by the pilot.

WBAS—Weather bureau airport station.

WCA—Wind correction angle.

WDLY—Widely.

WEA—Weather.

Weak link—A weak spot in the tow rope used in towing gliders into the air. Its breaking strength is specified by the FAA as a safety feature.

Weather advisory—Report of hazardous weather conditions not included in previous forecasts. See also *AIRMET* and *SIGMET*.

Weather broadcast—Aviation weather reported by FSSs for areas within approximately 150 miles of the station. It is broadcast hourly at 15 minutes past the hour.

Weathercock—The tendency of an aircraft to align itself lengthwise with the direction of the wind.

Weather minimums—The lowest amount of visibility and the lowest ceiling under which takeoff, landing or flight is allowed. Weather minimums vary between VFR and IFR, controlled and uncontrolled airspace, type of instrument approach and type of aircraft flying the approach.

Weather radar—Airborne radar which depicts areas of precipitation on the scope. Its most important feature is its ability to paint the areas within thunderstorms which are likely to be the most turbulent and thus to permit the pilot to plot a reasonably safe path through the storm.

Weather station—A service, usually at an airport and often co-located with an FSS, which collects weather information, and conducts pilot weather briefings.

Weathervane—The tendency of an aircraft on the ground to face into the wind, resulting from the wind hitting the vertical stabilizer.

Weight—The force of gravity which pulls an object to the center of the earth. It must be overcome by lift to achieve flight.

Weight and balance—The calculations involving the proper loading of an aircraft with reference to weight and the CG.

Westwind—A corporate jet manufactured by Israel Aircraft Industries and marketed in the U.S. by Atlantic Aviation.

Wipstall—See *vertical stall*.

Whirly bird—Slang for helicopter.

Whirly Girls, Inc.—See *aviation organizations*.

Whiskey—In aviation-radio phraseology the term used for the letter W.

Whiteout—The inability to distinguish ground from horizon and sky. A phenomenon occurring when flying over flat, featureless, snow-covered terrain when the sky is the same color as the ground.

Wicker basket—A gondola for ballooning, made of wicker.

Wilco—A phrase used in radio communication, meaning that the message has been received and will be complied with.

Westwind.

Winch launch—A means of launching a glider by using a ground-based winch rather than a towplane or automobile.

Wind—Air in horizontal motion caused by pressure or temperature differences in the atmosphere.

Wind, altimeter voice equipment—An automatic device which reports weather conditions with a computer-generated voice.

Wind arrow—A symbol used on weather maps to show wind direction and velocity. The shaft of the arrow shows the direction and each "feather" indicates 10 knots, each half "feather" five knots.

Wind-chill factor—The still-air temperature which would have the same cooling effect on exposed human flesh as a given combination of temperature and wind velocity.

Wind correction angle—The number of degrees divergent from the compass course which the pilot must steer in order to compensate for cross-wind components.

Wind drift—The movement of an airplane to the left or right of his line of flight, resulting from the force of the wind.

Windmilling propeller—In the event of engine stoppage, the tendency of the propeller to continue to turn because of the airstream moving across the blades. A windmilling propeller produces drag rather than thrust.

Winds aloft—Winds at altitudes above 1,500 feet agl. They are measured every six hours by radiosondes and reported in groups of figures representing true heading and velocity in knots. Winds aloft are not usually included in weather broadcasts, but pilots can request the information for whatever altitudes are of interest.

Winds-aloft chart—A weather map showing wind speed and direction (with wind arrows) at selected altitudes.

Windscreen—Windshield.

Wind shadow—An expression used primarily in soaring. It denotes an area of calm in the lee of windbreaks such as hills, buildings, trees. When it is sunny they are likely sources of thermals on a windy day.

Wind shear—An abrupt change in wind direction or velocity.

Wind shift—The veering or backing of the wind.

Windsock—A coneshaped cloth sleeve, usually orange or yellow, which catches the wind and points away from the direction from which it is blowing. Usually located on one or several prominent positions on the airport.

Wind tee—Tetrahedron.

Wind triangle—The basic concept of dead reckoning, involving true heading, airspeed, ground speed and wind direction and velocity. The calculations necessary to arrive at the ground speed and heading to make good the desired course can easily be accomplished with the slide-rule portion of a flight computer.

Wind velocity gradient—The horizontal wind shear close to the ground, cause by the frictional effect of the terrain.

Wing—The primary airfoil of an airplane, developing most of the needed lift.

Wing loading—The total forces exerted on a wing in flight, expressed in pounds per square foot of wing area. All aircraft have maximum wing-loading capability as part of the design concept.

Wingtip vortices—See *wake turbulence*.

WIP—Work in progress.

WK—Weak.

WKN—Weaken.

WND—Wind.

Wobble pump—An emergency fuel pump operated by manually pushing a handle back and forth.

Words twice—When used as a request it means that the communication is difficult to understand. Therefore, please say each phrase twice. Or it may be used to inform the listener that each phrase in the message will be spoken twice because of communication difficulties.

World Aeronautical Chart—An aeronatical chart in a scale of 1:1,000,000, or approximately 13.7 nm to the inch.

WP—Waypoint.

WR—Wet runway.

WRM—Warm.

WSR—Wet snow on runway.

WT—Weight.

Wulfsberg Electronics—Manufacturers of airborne telephone systems and other avionics. (11300 West 89th Street, Overland Park, KS 66214.)

WV—Wave.

WX—Weather.

X—Symbol for total obscuration (in sequence reports).
X—Xray (exray) (Phonetic alphabet).
XCVR—Transceiver.
XMTR—Transmitter.
XPDR—Transponder.
Xray—In aviation-radio phraseology the term used for the letter X.

Y—Yankee (phonetic alphabet).
Y—Yukon standard time.
Yankee—In aviation-radio phraseology the term used for the letter Y.
Yaw—The movement of the aircraft to either side, turning around its vertical axis (without banking).
Yaw axis—The vertical axis through the CG of the aircraft.
Yaw damper—A single-axis autopilot which automatically counteracts the tendency of some aircraft to yaw in flight.
Yaw string—A few inches of yarn attached inside the cockpit of gliders in view of the pilot. When it leans to one side or the other in flight, it indicates a slip or a skid and the need for correcting the degree of rudder pressure.
Yoke—Control wheel; stick

Z—Greenwich mean time.

Z—Zulu (phonetic alphabet).

Zero fuel weight—Some larger aircraft are limited for structural reasons in the amount of weight which can safely be carried when there is a minimum amount of fuel in the wing tanks. Since aircraft are likely to be low on fuel prior to landing, the zero fuel weight must be considered when on-loading passengers and /or freight.

Zero-zero—Phrase used to indicate that both ceiling and visibility are zero.

Zinc chromate primer—A primer used on bare aluminum and magnesium to prevent corrosion.

ZL—Freezing drizzle (in sequence reports).

ZM—Z-marker.

Z-marker—A VHF radio beacon broadcasting straight up at a frequency of 75 MHz. It is used to obtain position information.

Zoom—A means of gaining altitude by increasing airspeed and then pulling back on the control wheel and climbing sharply until the excessive momentum is lost and the aircraft approaches stall.

ZR—Freezing rain (in sequence reports).

Zulu—In aviation-radio phraseology the term used for the latter Z.

Zulu Time—Greenwich Mean Time.